Topics in
ELECTROMAGNETIC WAVES
Devices, Effects and Applications

Topics in
ELECTROMAGNETIC WAVES
Devices, Effects and Applications

Edited by
Jitendra Behari

Anamaya Publishers
New Delhi

Edited by
Jitendra Behari
Dean, School of Environmental Sciences
Jawaharlal Nehru University
New Delhi - 110 067, India

Copyright © 2005 Anamaya Publishers

ANAMAYA PUBLISHERS
F-230, Lado Sarai, New Delhi - 110 030, India
e-mail: anamayapub@vsnl.net
        anamayapub@indiatimes.com

ISBN 81-88342-10-6

Published by Manish Sejwal for Anamaya Publishers,
F-230, Lado Sarai, New Delhi - 110 030. Printed in India.

Preface

The present volume is based on the papers presented at the conference on "Microwave Applications: Medicine, Agriculture, Remote Sensing and Industry" held at Solid State Physics Laboratory, Delhi, under the auspices of Microwave Applications Society of India (MASI). It was sponsored by Defence Research and Development Organization, Indian Council of Medical Research and Jawaharlal Nehru University, New Delhi. Financial assistance for publication of this book from Department of Science and Technology, New Delhi, is thankfully acknowledged. The conference was aimed at providing a common platform to researchers in the area of electromagnetic field related applications to deliberate and establishing an interface for providing directions to future endeavors.

The book is divided broadly into four parts: the first part deals with the recent advances in the development of microwave devices, the second with another important but not so well understood aspect: namely, their biological effects, their basis and applications. A much talked about issue relating to mobile phone safety is also covered. While Microwaves have found its way in industry, its domestic applications other than microwave ovens is still not appreciated. It is partly attempted in the third section, where its applications in agriculture as Time Domain Reflectometry methods is also discussed. Laser and millimeters-waves are now finding greater applications in almost every sphere of human activity: Environmental and therapeutical aspects etc. other articles are included so as to be of interest to a wide variety of readers. Use of millimeter waves for communication is now finding greater acceptability. These titles are aimed at providing state-of-the-art information on the concerned subject.

J. Behari

Contents

Topics in Electromagnetic Waves: Devices, Effects and Applications
Edited by J. Behari
Copyright © 2005, Anamaya Publishers, New Delhi, India

1. Device Design of AlGaAs/GaAs Abrupt Junction Heterostructure Bipolar Transistor

Vivan B. Narang and S.R. Shukla

Solid State Physics Laboratory, Delhi-110054, India

Abstract: This article discusses the design considerations of AlGaAs/GaAs heterojunction bipolar transistor based on thermionic-field-diffusion model. The dc current gain has been studied as a function of emitter and base dopings as well as their widths. The cut-off frequency has been simulated for various current densities. The design parameters of the device have been determined.

Introduction

AlGaAs/GaAs heterostructure bipolar transistor (HBT), by virtue of its structure, have many advantages for high speed operation at microwave and mm-wave frequencies. It has demonstrated excellent potential as a power device due to its high current carrying capability. In comparison with Si bipolar transistor, the AlGaAs/GaAs HBT have reduced base resistance, lower base-emitter capacitance resulting in higher cut-off frequency and have higher Early voltage leading to better linearity.

Device design plays a crucial role in the fabrication of integrated circuits and devices. Repeated fabrication process involves a lot of efforts and is not economical. Apart from this, there are many parameters, which affect the device performance and, thus, the hit and trial method does not work at high frequencies. Thus, device design plays a very important role in achieving the device performance as per required specification. Knowledge of parameters such as current gain β, cut-off frequency f_T and breakdown voltage is an essential requirement for designing HBT device. The physical parameters of the device such as doping density and widths of emitter, base and collector are used as input parameters in this process.

The theoretical work has been done in this direction in the literature. Lundstorm [1] has developed an Ebers Moll model based on thermionic emission for uniform and graded base HBT. But this model does not include the effect of electron tunnelling across the conduction band spike and hence, is unable to describe the current transport in abrupt junction HBT. Grinberg et al. [2] have developed thermionic-field-diffusion model for AlGaAs/GaAs HBT. However, they have not taken the effect of parasitic resistance into account.

In this article, the device design of AlGaAs/GaAs heterostructure bipolar transistor has been presented which is based on the simulation performed using thermionic-field-diffusion model which includes high current effects such as base push out and parasitic resistance and, at the same time, gives a consistent set of equations to evaluate collector current characteristics, base current characteristics, current gain and cut-off frequency.

Theoretical Approach

To design a device for required parameters like current gain β, cut-off frequency f_T and breakdown voltage using physical parameters of the device, we require simulation of these parameters from device physical parameters based on some appropriate model. For instance, to simulate the current gain β for various base-emitter voltage, the collector current and base current characteristics (Gummel plot) are needed. This has been accomplished using thermionic-field-diffusion (TFD) model including the effects of base push-out and parasitic resistance in this work [5, 6]. The cut-off frequency vs collector current density requires the computation of various delay times, namely, emitter charging time τ_E, base transit time τ_{BT}, collector signal delay time τ'_{CT} and collector charging time τ_C. All these delay times have been computed. The breakdown voltage is obtained from the collector doping density and critical electrical field.

The variation of collector current I_C, base current I_B and current gain β as a function of doping concentrations (N_E, N_B and N_C) and their widths (W_E, W_B and W_C) have been studied. The initial parameters for the device design are listed in Table 1. The x in $Al_xGa_{1-x}As$ emitter is equal to 0.3 and the junction area is $7 \times 7 \ \mu m^2$.

Table 1. HBT Device Structure

	Doping Density per cm^3	Width (Å)
Emitter	5×10^{17}	1000
Base	1×10^{19}	1000
Collector	5×10^{16}	3000

Results And Discussion

The variation of current gain β as a function of emitter and base doping density as well as their widths, considering one function at a time and keeping all the other device parameters constant, are studied and shown in Figs. 1 and 2. It is seen that as we increase N_E, the base current decreases due to increase in potential barrier resulting in increase in β. But increase in emitter doping also results in enhancing the base-emitter capacitance, which leads to lower the cut-off frequency. Thus, an intermediate value of N_E as 5×10^{17} /cm^3 is taken for device design. Further, it is seen that as base doping increases, the current gain decreases. However, large base doping results in smaller base resistance leading to high cut-off frequency. Moreover, this also avoids punch through as we consider the smaller base-width to increase base transit time. Thus, the base doping of 5×10^{19}/cm^3 is chosen.

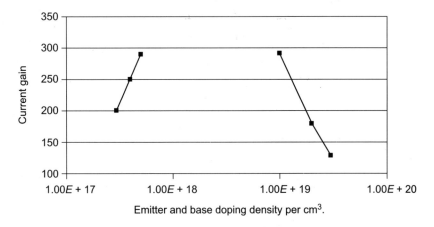

Fig. 1   Variation of current gain with emitter and base doping density.

It is evident from Fig. 2 that the current gain increases with the increase in emitter width on account of decrease in base current which is due to decrease in hole injection current. But, increase in W_E also leads to increase in minority carrier storage time. Thus, a medium value of $W_E \sim 1000\text{-}1200$ Å is taken. It is seen that the current gain decreases with the increase in base width. This results from the increase of quasi-neutral base recombination current with increase in base width. Moreover, reducing W_B will result in punch through of the base.

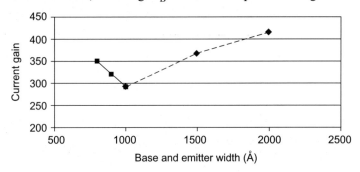

Fig. 2   Variation of current gain with base and emitter widths.

Thus, the base width $W_B \sim 800\text{-}1000$ Å is taken. It has been found that the current remains unchanged if we vary the collector doping or the collector width. However, the breakdown voltage depends upon collector doping. Therefore, for breakdown voltage ~18 V, we have taken collector doping $\sim 5 \times 10^{16}/\text{cm}^3$ and collector width ~ 3000-3500 Å. The cut-off frequency f_T depends upon collector current density J_c and, hence, upon the applied base-emitter voltage V_{be}. The cut-off frequency is simulated including all the four delay times mentioned above and is plotted as a function of J_c in Fig. 3.

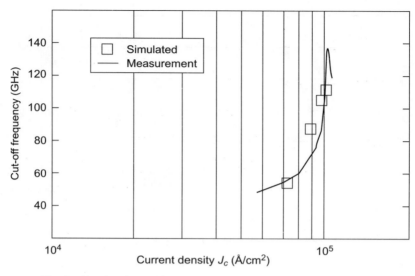

Fig. 3   Simulated cut-off frequency compared with measured data [3].

Conclusions

In this article, the device design of AlGaAs/GaAs abrupt junction heterojunction bipolar transistor has been presented which is based on the simulation performed using thermionic-field-diffusion model. The present model gives simulated results for $N/p^+/n$ AlGaAs/GaAs HBT which are in good agreement with the published data [3, 4]. The design parameters were determined for the specified values of current gain $\beta \sim 250\text{-}300$, $f_T \sim 90\text{-}100$ GHz, and breakdown voltage ~ 18 V.

References

1. Lundstorm, M.S. *Solid St. Electron.*, **29**, 1986, p. 1173.
2. Grinberg, A.A., M.S. Shur, R.J. Fischer and Morkoc, H. *IEEE Trans. Electron Devices*, **ED-31**, 1984, p. 1758.
3. Welsor, R.E., Pan, N., Lutz, C.R., Vu, D.P., Zampardi, P.J., Pierson, R.L. and Mcdermott, B.T. "GaAs Heterojunction Bipolar Transistor Emitter Design" GaAS Symp. Digest 1999.
4. Liu, W. *Electronic Letters*, **27**, No.23, 1991, p. 2115.
5. Shukla, S.R., Narang, V.B., Singh, D.B. and Vyas, H.P. Proc. XI Intl. Workshop Phys. Semiconductor Dev., Dec. 2001, Delhi.
6. Narang, V.B., Shukla, S.R. and Vyas, H.P. Proc. XII Intl. Workshop Phys. Semiconductor Dev., Dec. 2003, p. 839, Delhi.

Topics in Electromagnetic Waves: Devices, Effects and Applications
Edited by J. Behari

2. Optimization of Various Parameters for Compact Antenna with Broadband Operation

R.M. Vani*, S.F. Farida** and P.V. Hunagund*

*University Science Instrumentation Centre, Department of Applied Electronics, Gulbarga University, Gulbarga-585 106, Karnataka, India

**Department of Electrical Engineering, Salt Lake Community College, UTAH-84130, USA

Abstract: This article presents the typical configuration of narrow vertical slots on the patch with gap coupling used to achieve compact broadband operation. The various parameters have been optimized to obtain broader bandwidth without much increase in area.

Introduction

A narrow bandwidth is a major disadvantage of microstrip antennas in practical applications. For present day wireless communication systems, the required operating bandwidths for antenna are about 7.6% for a global system for mobile communication (GSM;890-960 MHz), 9.5% for a digital communication system (DCS: 1710-1880 MHz) 7.5% for a personal communication system (PCS; 1850-1990 MHz) and 12.2% for a universal mobile telecommunication system (UMTS; 1920-2170 MHz). To meet these bandwidth requirements, many bandwidth enhancement or broadband techniques for microstrip antennas have been reported recently [1-4]. This article presents the typical configuration of narrow vertical slots at $L/4$ distance on the patch surface with gap coupling and discusses optimization of various parameters like dimensions of driven and parasitic patches, gap between the two patches, number of slots, length of slots, position of slots, feed point location etc. The zeland IE3D simulation software has been employed to get impedance matching and to accurately predict the electrical characteristics of the designed patches.

Antenna Design and Experimental Results

The patches are designed by using glass epoxy material with $\varepsilon_r = 4.3$, $h = 1.6$ mm, length $L = 30$ mm and width $W = 20$ mm. In this technique the two patches have been parasitically coupled and only one patch is fed with coaxial probe feed. Initially the two patches were designed with single slot and gap coupling. The gap between the parasitic and driven elements S has been simulated and optimized using IE3D software by varying the value from 0.025 λ_g to 0.01 λ_g, where λ_g is the effective guide wavelength. It has been observed that when $S = 0.025\lambda_g$, then maximum possible bandwidth can be obtained.

By maintaining this gap, next the dimension of parasitic patch is varied with respect to driven patch and it is found that the maximum bandwidth can be obtained only when the size of parasitic patch is same as that of driven patch.

Then by keeping above parameters constant, the number of slots on the patches were increased from 1 to 3 slots and it has been observed the maxi frequency reduction can be obtained by increasing number of slots. The variation in the frequency is also observed by varying the length of slots and by changing the position of slots on the patch surface. If the slots are moved towards the centre, the frequency reduction is more, but the bandwidth is reduced. To get maximum possible bandwidth slots are equally spaced at $L/4$ distance on the patch surface.

With all optimized parameters, the prototypes of the proposed antenna have been designed starting with single slot to three slots and arrangements of slots with two gap coupled patches are as shown in Fig. 1. The slots are embedded at $L/4$ distance all having same length $l_s = 18$ mm and width $w_s = 1$ mm, respectively. It has been observed that if the number of slots on the patch are changed from 1 to 3 then the frequency goes on reducing with same bandwidth. The results for all the configurations are summarized in Table 1. Fig. 2 shows the return loss characteristics for all the configurations. The radiation pattern of all antennas remains broadside. Fig. 3 shows the radiation pattern in the E-plane and H-plane for the gap-coupled slotted RMSA with three slots.

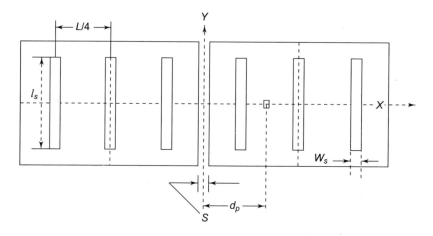

Fig. 1   Geometry of gap-coupled slotted RMSA with three slots ($\varepsilon_r = 4.3$, $h = 1.76$ mm, $L = 30$ mm, $W = 20$ mm, $l_s = 18$ mm, $w_s = 1$ mm, $S = 1.585$ mm, $r_p = 0.6$ mm).

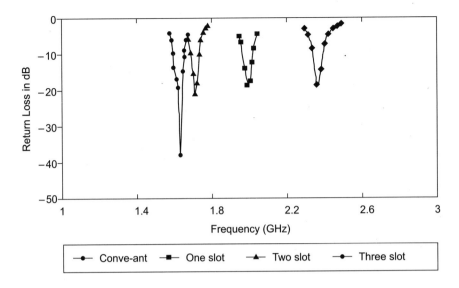

Fig. 2   Return loss characteristics of proposed antennas and conventional antenna.

Table 1.   Comparison of gap-coupled slotted RMSA configurations

Prototype antennas	Feed from center axis d_p (mm)	Frequency (GHz)	VSWR	Return loss (dB)	Bandwidth (MHz, %age)	
Conventional RMSA	5.00	2.367	1.259	−18.81	50	2.11
Gap-coupled slotted RMSA (each one slot)	11.00	1.994	1.251	−19.04	52	2.60
Gap-coupled slotted RMSA (each two slots)	21.50	1.713	1.188	−21.30	54	3.15
Gap-coupled slotted RMSA (each three slots)	11.00	1.633	1.025	−38.15	52	3.18

Conclusion

Various configurations of gap-coupled slotted RMSA's have been designed and optimization of various parameters for compact and broadband operation has been studied. Further, it has been observed that the frequency reduction is more with increase in number of slots on the patch, which leads to more size reduction. Moreover, by putting parasitic patch with the driven patch, broader bandwidth has been realized.

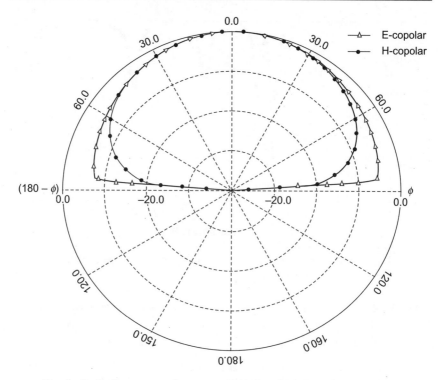

Fig. 3 Radiation patterns for gap-coupled slotted RMSA with three slots.

References

1. Wong, K.L. and Hsu, W.H. A broadband rectangular patch antenna with a pair of wide slits, *IEEE Trans. Antennas Propagat*, **AP-49,** (2001), 1345-1347.
2. Deshmukh, A.A. and Kumar, Girish, Compact broadband gap-coupled square ring and W-shaped microstrip antennas, Proc. of IRSI, (2001), 853-861.
3. Wu, C.K. and Wong, K.L. "Broadband microstrip antenna with driectly coupled and gap-coupled parasitic patches", *Microwave Opt. Technol. Lett.*, **22,** Sept. 1999, pp. 348-349.
4. Wong, K.L., Au., T.M., Luk, K.M. and Lee, K.F. "Two-layer five-patch broadband microstrip antenna", *Electron Lett.*, **31,** Sept. 1995, pp. 1621-1622.

Topics in Electromagnetic Waves: Devices, Effects and Applications
Edited by J. Behari
Copyright © 2005, Anamaya Publishers, New Delhi, India

3. Design of a 0.75-1.5 GHz MESFET MMIC Amplifier

Vijesh Arora and S.R. Shukla

Solid State Physics Laboratory, Delhi-110 054, India

Abstract: This article describes the design and performance of a two-stage amplifier operating in the 0.75-1.5 GHz frequency range. The MMIC amplifier has a gain of 20 ± 0.5dB, noise-figure less than 3 dB and power output (1 dB gain compression) exceeding 17 dBm. The chip was designed using 0.7 μm gate length MESFET devices, having a gate-width of 900μm to achieve the design goals. The emphasis in the design was to account for process spreads and ensure first-pass success, which is evident from the measured results.

Introduction

MMIC amplifiers are frequently used as gain blocks for providing high gain, moderate noise figure and reasonable power output. The various applications of gain blocks are driver amplifiers for medium power amplifiers, buffer amplifiers, loop-amplifier for oscillators and cardinal amplifiers in low power applications.

In this article, the design and performance of a two-stage MMIC feedback amplifier is described. This amplifier provides high gain with moderate noise figure and power output.

Amplifier Design

The resistive shunt feedback approach was followed for this design as it imparts gain flatness, improves the input and output return losses, provides reasonable noise and power performance and ensures that the circuit performance is fairly insensitive to process variations [1].

The initial design of the shunt feedback amplifier was carried out using a simplified device model. In this model, the device is represented as an ideal voltage controlled current source. The feedback resistance R_{fb} is connected between the gate and drain terminals.

For an input/output VSWR of $K : 1$,

$$R_{fb} = K_{gm} Z_0^2 + (K-1) Z_0 \text{ and } G = 2(K_{gm}Z_0 - 1)/(1 + K)$$

For this design, $K = 1.5$ and $G = 3.2$, which necessitates a device with g_m exceeding 66 mS.

For selecting a suitable device, it was essential to obtain the equivalent circuit parameters for 6×150 μm and 8×150 μm MESFET.

The equivalent circuit element values of a device with gate width W were scaled from a 4×150 μm device, which was rigorously modeled after S-parameter and dc measurements. The scaling equations are:

$$\text{Scaling factor } n = W \text{ (mm)}/0.6$$

$$C_{gsw} = n.C_{gs}, \quad C_{gdw} = n.C_{gd}, \quad C_{dsw} = n.C_{ds}; \quad g_{mw} = n.g_m;$$
$$R_{gsdw} \sim R_{gs}\,d/n, \quad R_{dsw} = R_{ds}/n \text{ and } L_{gsdw} \sim L_{gs}\,d/n$$

The gain and return loss (at the center frequency) of 600, 900 and 1200 μm devices along with the respective feedback resistances are plotted in Fig. 1. In order to meet the specifications, the design was based on 900 μm devices biased at 5 V and 0.5 I_{dss}. The circuit simulations were carried out using HP-Eesofs: Series IV software.

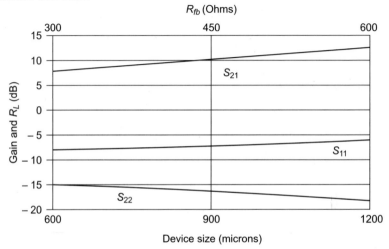

Fig. 1   Device size (microns).

The noise parameters F_{min}, R_n and Γ_{opt} were scaled from the 600 μm noise data and small-signal equivalent parameters using a modified Podell noise model [2]. The noise model equations utilized in the design were

$$F_{min} = 1 + 4\pi f C_{gswv}(K_{oe}K_b R_{bw}/g_{mw}), \text{ where } K_b = I_{ds}/I_{dss}, \quad K_0 = R_n\,(g_m/eK_b)$$

and, R_n is the measured noise resistance with K_0 a foundry specific constant. Also,

$$R_{bw} = R_{gw} + R_{sw} + 4K_b^2\,R_i$$

The frequency dependent noise resistance, described by an empirical expression [3], is: $R_n = R_{no} + 0.5\sqrt{f} - 0.81f$ (at low frequencies, $R_{no} \sim R_n$).

Γ_{opt} comprises of g_{sopt} and b_{sopt} described as

$$G_{sopt} = g_1\sqrt{(1 + A^{-1})}; \quad g_1 = (2\pi f C_{gsw})^2\,R_{bw} \text{ and } A = (F_{min}-1)^2/4F_{min}$$

F_{min} is the measured minimum noise factor. Also, $B_{sopt} = -2\pi f C_{gsw}$.

The power output at 1dB compression was estimated using the linear approximations. The power capability of the G7A process is approximately 100 mW/mm. Therefore, the predicted output power (1dB gain compression) for a 900 μm device, assuming a loss of 1 dB in the output matching, is 18.5 dBm.

Two-stage design was then carried out to deliver a gain of 20 + 0.5 dB with input return loss better than 10 dB, output return loss better than 12 dB, NF less than 3 dB and power output (at 1dB gain compression) exceeding 17 dBm.

The chip size was 2.2 mm × 2.0 mm and is depicted in Fig. 2.

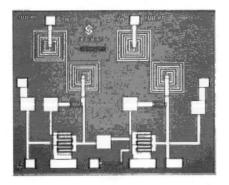

Fig. 2   MMIC chip.

Measurements

The amplifiers were fabricated, diced and mounted onto carriers for RF measurements. Typical measured results illustrated in Figs. 3 and 4, conform to

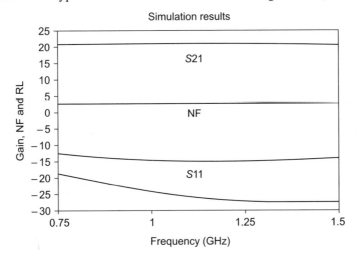

Fig. 3   Amplifier performance.

the simulations and demonstrate that all design specifications were successfully met in the first foundry run.

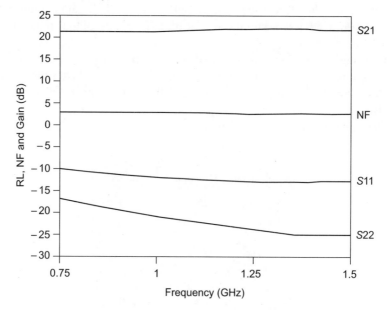

Fig. 4   Measured gain, return loss and NF.

Conclusion

A two-stage MMIC amplifier has been successfully designed and tested. The measured performance is in good agreement with the circuit simulations. All the design specifications were met in the first foundry run.150 amplifier chips (21106) have been delivered to the users. This gain block operating in the P and L bands is a standard product available from GAETEC foundry.

References

1. K.B. Niclas et al. "The Matched Feedback Amplifier: Ultra Wide-band Amplification with GaAs MESFET's", *IEEE Trans. Microwaves Theory & Tech.*, **MTT-28**, 285–294, April 1980.

2. A.F. Podell, "A Functional GaAs FET Noise Model", *IEEE Trans. Electron Devices*, **ED-28**, 511–517, May 1981.

3. V. Arora et al. " MMIC Design Oriented GaAs FET Noise Model", Proc. XIIth Intl. Workshop, Physics of Semiconductor Devices, 582–584, December 2003.

Topics in Electromagnetic Waves: Devices, Effects and Applications
Edited by J. Behari

4. Performance Comparison Between HMCE and IMCE Serrated CATRs

P. Siddaiah, P.V. Subbaiah* and T. Venkata Rama Krishna

Department of E.C.E, R.V.R. & J.C College of Engineering, Chowavaram, Guntur, India

*Department of E.C.E, V.R. Siddhartha Engineering College, Vijayawada, India

Abstract: The Compact Antenna Test Range (CATR) provides uniform illumination within the Fresnel region to test the antenna. Application of serrated edges has been shown to be a good method to control diffraction at the edges of the reflectors. The performance of several simple serrations have been analyzed earlier. In this paper the Fresnel fields of a Hybrid Modulated Complex Exponential (HMCE) serrated CATR and Identical Modulated Complex Exponential (IMCE) serrated CATR are analyzed by using Physical Optics technique. It is observed that HMCE serrated CATR gives, less ripple and enhanced quiet zone width than IMCE.

Introduction

A professional testing system should provide for accuracy, reliable and repeatable performance in addition to possession of the capabilities for handling several antenna performance criteria, assurance of field free region, with known degree of external interference, if any, and maintainability of test environment. The parameters that are normally to be assessed include: radiation pattern, beamwidth, gain, directivity, front to back ratio, polarization etc. The testing of microwave antennas usually requires that the antenna under test be illuminated by a uniform plane wave. The performance of open test ranges may not be consistent due to changes in weather and extraneous interfering sources. The Microwave Anechoic Chamber (MAC) provides immunity to weather and EMI, but only small antennas operated at moderate frequencies can be measured in a MAC due to the need to respect the Fraunhofer criterion. In the Near Field-Far Field ranges (NF-FF), the NF data can be measured in a MAC and FF can be computed. NF-FF technique does not facilitate a real time measurement [1].

The Compact Antenna Test Ranges sets up a plane wave illumination by employing a collimated beam from a parabolic reflector, obviating the need to respect Fraunhofer criterion. This aspect enables compact size of the range and qualifies the CATR to be a wide band facility. This is an abrupt discontinuity of the reflected field at the surface termination. The diffracted signal emanating from the surface termination causes amplitude and phase variations of the field that illuminates the test antenna.

Serrated edge termination is intended to reduce the diffraction problem. The number, shape and height and width of the serrations are chosen such that the diffracted field emanating from these serrations will mutually get cancelled among themselves. In this article, aperture formulation of the physical optics (PO) technique is used to predict the Fresnel fields of a square aperture reflector with: (a) Hybrid Modulated Complex Exponential (HMCE) and (b) Identical Modulated Complex Exponential (IMCE) serrations are evaluated.

Method of Analysis

Performance comparison is made based on two criteria i.e. size of the quiet zone and amount of ripple in the quiet zone. A square aperture reflector of $45\lambda \times 45\lambda$ is equipped with HMCE and IMCE serrations as shown in Figs. 1 and 2, respectively. To compare the quiet zone performance of a plane square aperture reflector employing two different types of serrations viz. (a) width modulated complex exponential and width and height modulated complex exponential (HMCE) (b) width and height modulated complex exponential (IMCE).

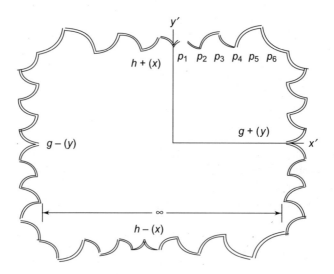

Fig. 1   Square aperture reflector with HMCE serrations.

It is very difficult to field an analytical expression in the closed form for the Fresnel zone field of an aperture with serrated edges. Hence, it is better to decompose the aperture area S into three parts S_1, S_2 and S_3 such that $S = S_1 + S_2 - S_3$ and apply a quasi-analytical expression to derive the Fresnel field [2-3].

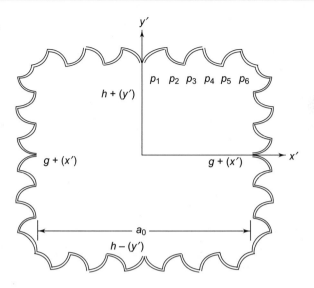

Fig. 2 Square aperture reflector with IMCE serrations.

The HMCE serration described by the boundary functions $g^+(y')$ and $g^-(y')$ are expressed as Fourier series of width modulated complex exponential with rate of rise a_i and $h^+(x')$ and $h^-(x')$ are described as the Fourier series of width and height modulated complex exponential serrations.

In IMCE, the serrations described by the boundary functions $g^+(y'), g^-(y')$ and $h^+(x'), h^-(x')$ are expressed as Fourier series of width and height modulated complex exponential serrations [4-10].

HMCE Serrations

Fourier Series of Width Modulated Complex Exponential Serrations

$$g^+(y') = \frac{a_0}{2} + \frac{1}{p_6} = \left[t\left(p_1 + \frac{1}{a_1}\left(e^{-a_1 p_1} - 1\right) + \frac{1}{a_2}\left(1 - e^{-a_2(p_2-p_1)}\right)\right) \right.$$

$$+ t_2\left(p_3 - p_2\right) + \frac{1}{a_3}\left(e^{-a_3(p_3-p_2)} - 1\right) + \frac{1}{a_4}\left(\left(1 - e^{-a_4(p_4-p_3)} - 1\right)\right)$$

$$\left. + t\left(p_5 - p_4\right) + \frac{1}{a_5}\left(e^{-a_5(p_5-p_4)} - 1\right) + \frac{1}{a_6}\left(1 - e^{-a_6(p_6-p_5)}\right) \right]$$

$$+ \sum_{n=1}^{\infty} \frac{1}{p_6}\left[t\left(\frac{1}{q}\sin q_1 - b_1\right)\left(e^{-a_1 p_1}\left(-a_1 \cos q_1 + a_2 \cos q_1\right) - a_1\right)\right.$$

$$\left. + b_2\left(e^{-a_2 p_2}\left(-a_2 \cos q_2 + q \sin q_2\right) + \left(e^{-a_2 p_1}\left(a_2 \cos q_1 - q \sin q_1\right)\right)\right) \right]$$

$$+\frac{1}{6}\left[t\left(\frac{1}{q}(\sin q_3 - \sin q_2) - b_3 e^{a_3 p_2}\left(-a_3 e^{-a_3 p_3}\cos q_3 + e^{-a_3 p_3} q \sin q_3\right.\right.\right.$$

$$+a_3 e^{-a_3 p_2}\cos q_2 - e^{-a_3 p_2}\sin q_2$$

$$\left.\left.\left.+b_4 e^{a_4 p_3}\left(e^{-a_4 p_4}(-a_4 \cos q_4 + q \sin q_4) + e^{-a_4 p_3}(a_4 \cos q_3 - q \sin q_3)\right)\right)\right]\right.$$

$$+\frac{1}{p_6}\left[t\left(\frac{1}{q}(\sin q_5 - \sin q_4) - b_5\left(-a_5 \cos q_5\, e^{-a_5 p_5} + e^{-a_5 p_5} q \sin q_5\right.\right.\right.$$

$$\left.+a_5 e^{-a_5 p_4}\cos q_4 - e^{-a_5 p_4} q \sin q_4\right) + b_6 e^{a_6 p_5}\left(-a_6 e^{-a_6 p_6}\cos q_6\right)$$

$$\left.\left.\left.+q e^{-a_6 p_6}\sin q_6 + a_6 e^{-a_6 p_5}\cos q_5 - q e^{-a_6 p_5}\sin q_5\right)\right)\right]\cos(qy')$$

$$q = n\pi/p_6,\ q_i = q p_i,\ b_i = 1/(a_i^2 + q^2)$$

Fourier Series of Width and Height Modulated Complex Exponential Serrations

$$h^+(x') = \frac{a_0}{2} + \frac{1}{p_6}\left[t_1\left(p_1 + \frac{1}{a_1}\left(e^{-a_1 p_1} - 1\right) + \frac{1}{a_2}\left(1 - e^{-a_2(p_2 - p_1)}\right)\right) + \left(t_2(p_3 - p_2)\right.\right.$$

$$+\frac{1}{a_3}\left(e^{-a_3(p_3 - p_2)} - 1\right) + \frac{1}{a_4}\left(1 - e^{-a_4(p_4 - p_3)} - 1\right)\right) + \left(t_3(p_5 - p_4)\right.$$

$$\left.\left.+\frac{1}{a_5}\left(e^{-a_5(p_5 - p_4)} - 1\right) + \frac{1}{a_6}\left(1 - e^{-a_6(p_6 - p_5)}\right)\right)\right]$$

$$+\sum_{n=1}^{\infty}\frac{1}{p_6}\left[t_1\left(\frac{1}{q}\sin q_1 - b_1\left(e^{-a_1 p_1}(a_1 \cos q_1 + a_2 \cos q_1) - a_1\right)\right.\right.$$

$$\left.\left.+b_2\left(e^{-a_2 p_2}(-a_2 \cos q_2 + q \sin q_2)\right) + \left(e^{-a_2 p_1}(a_2 \cos q_1 - q \sin q_1)\right)\right)\right]$$

$$+\frac{1}{p_6}\left[t_2\left(\frac{1}{q}(\sin q_3 - \sin q_2) - b_3 e^{a_3 p_2}\left(-a_3 e^{-a_3 p_3}\cos q_3 + e^{-a_3 p_3} q \sin q_3\right.\right.\right.$$

$$\left.+a_3 e^{-a_3 p_2}\cos q_2 - e^{-a_3 p_2} q \sin q_2\right) + b_4 e^{a_4 p_3}\left(e^{-a_4 p_4}(-a_4 \cos q_4 + q \sin q_4)\right.$$

$$\left.\left.\left.+e^{-a_4 p_3}(a_4 \cos q_3 - q \sin q_3)\right)\right] + \frac{1}{p_6}\left[t_3\left(\frac{1}{q}(\sin q_5 - \sin q_4)\right.\right.$$

$$-b_5\left(-a_5 \cos q_5 e^{-a_5 p_5} + e^{-a_5 p_5} q \sin q_5 + a_5 e^{-a_5 p_4}\cos q_4 - e^{-a_5 p_4} q \sin q_4\right) + b_6 e^{a_6 p_5}$$

$$\left.\left.\times\left(-a_6 e^{-a_6 p_6}\cos q_6 + q e^{-a_6 p_6}\sin q_6 + a_6 e^{-a_6 p_5}\cos q_5 - q e^{-a_6 p_5}\sin q_5\right)\right)\right]\cos(qy')$$

$$q = n\pi/p_6,\ q_i = q p_i,\ b_i = 1/(a_i^2 + q^2)$$

IMCE Serrations

Fourier Series of Width and Height Modulated Complex Exponential Serrations

$$g^+(y') = \frac{a_0}{2} + \frac{1}{p_6}\left[t_1\left(p_1 + \frac{1}{a_1}\left(e^{-a_1 p_1} - 1\right) + \frac{1}{a_2}\left(1 - e^{-a_2(p_2 - p_1)}\right)\right)\right.$$

$$+ t_2\left((p_3 - p_2) + \frac{1}{a_3}\left(e^{-a_3(p_3 - p_2)} - 1\right) + \frac{1}{a_4}\left(1 - e^{-a_4(p_4 - p_3)} - 1\right)\right)$$

$$\left.+ t_3\left((p_5 - p_4) + \frac{1}{a_5}\left(e^{-a_5(p_5 - p_4)} - 1\right) + \frac{1}{a_6}\left(1 - e^{-a_6(p_6 - p_5)}\right)\right)\right]$$

$$+ \sum_{n=1}^{\infty} \frac{1}{p_6}\left[t_1\left(\frac{1}{q}\sin q_1 - b_1\left(e^{-a_1 p_1}(a_1\cos q_1 + a_2\cos q_1) - a_1\right)\right.\right.$$

$$\left.+ b_2\left(e^{-a_2 p_2}(-a_2\cos q_2 + q\sin q_2)\right) + \left(e^{-a_2 p_1}(a_2\cos q_1 - q\sin q_1)\right)\right)\right]$$

$$+ \frac{1}{p_6}\left[t_2\left(\frac{1}{q}(\sin q_3 - \sin q_2) - b_3 e^{a_3 p_2}\left(-a_3 e^{-a_3 p_3}\cos q_3 + e^{-a_3 p_3}q\sin q_3\right.\right.\right.$$

$$+ a_3 e^{-a_3 p_2}\cos q_2 - e^{-a_3 p_2}q\sin q_2\right) + b_4 e^{a_4 p_3}\left(e^{-a_4 p_4}(-a_4\cos q_4 + q\sin q_4)\right.$$

$$\left.\left.+ e^{-a_4 p_3}(a_4\cos q_3 - q\sin q_3)\right)\right] + \frac{1}{p_6}\left[t_3\left(\frac{1}{q}(\sin q_5 - \sin q_4)\right.\right.$$

$$- b_5\left(-a_5\cos q_5\, e^{-a_5 p_5} + e^{-a_5 p_5}q\sin q_5 + a_5 e^{-a_5 p_4}\cos q_4 - e^{-a_5 p_4}q\sin q_4\right)$$

$$+ b_6 e^{e_6 p_5}\left(-a_6 e^{-a_6 p_6}\cos q_6 + q e^{-a_6 p_6}\sin q_6 + a_6 e^{-a_6 p_5}\cos q_5\right.$$

$$\left.\left.\left.- q e^{-a_6 p_5}\sin q_5\right)\right)\right]\right] \cos(qy')$$

$$q = n\pi/p_6,\ q_i = q p_i,\ b_i = 1/(a^2_i + q^2)$$

The above Fourier series representations in conjunction with analytical expression of Beeckman [2] gives the Fresnel zone field.

Results

A square aperture of dimension $45\lambda \times 45\lambda$ is equipped with HMCE and IMCE serrations are shown in Figs. 1 and 2, respectively. Fresenel field calculations are made at the distance of 64λ along the z-axis. The variation of relative power in dB with transverse distance in wavelengths is furnished in Figs. 3 and 4. From these figures, it is observed that by proper selection of width and height modulation factors (Tables 1 and 2), lesser ripple and enhanced quiet zone width are observed in HMCE than IMCE. It is concluded that, hybrid type of serrated CATR gives better performance than IMCE.

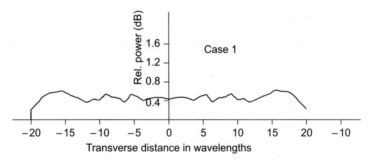

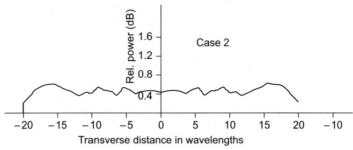

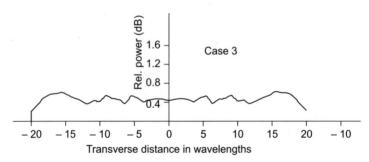

Fig. 3   Fresnel zone field of $45\lambda \times 45\lambda$ HMCE reflector for cases 1, 2 and 3.

Table 1.   Height Modulation Factors

Case	P	P_1/P	P_2/P	P_3/P	P_4/P	P_5/P	P_6/P
1	$(a_0/2)/30$	2.00	4.00	8.00	12.00	24.00	30.00
2	$(a_0/2)/22.5$	2.40	3.20	8.00	09.60	19.20	22.50
3	$(a_0/2)/15$	2.30	3.33	6.93	08.66	13.33	15.00

Table 2.   Height Modulation Factors

t	t_1/t	t_2/t	t_3/t
5λ	1	1.5	2

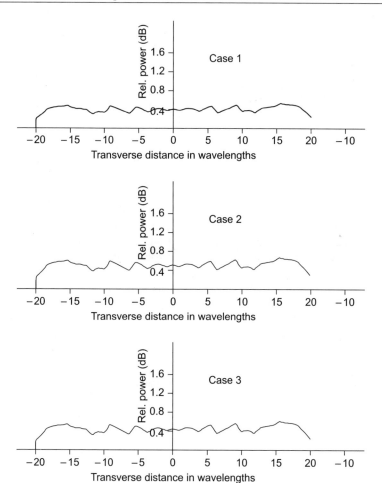

Fig. 4   Fresnel zone field of $45\lambda \times 45\lambda$ IMCE reflector for cases 1, 2 and 3.

Acknowledgements

The authors place on record their grateful thanks to the authorities of the R.V.R. & J.C. College of Engineering, Guntur and V.R. Siddhartha Engineering College, Vijayawada for providing the facilities.

References

1. Gary E. Evans. 'Antenna Measurement Techniques'. Artech House Incorporated 1990.

2. P.A. Beeckman. 'Prediction of the Fresnel Field of a Compact Antenna Test Range With Serrated Edges', IEE Proceedings on Microwave Antennas and Propagation, vol. 133, pt H, no. 2, April 1986, p. 108–114.

3. M.C. Chandra Mouly, P.V. Subbaiah, A.V.V. Satyanarayana and K. Madhavi, 'Analysis of Serrated Edge Reflectors Employed in Compact Antenna Test Ranges', Discussion Meeting on Numerical Techniques in Antenna Design, Indian Institute of Science, Bangalore, December 17–19, 1992, p. 18–23.

4. P. Siddaiah, and P.V. Subbaiah, 'Performance of Compact Antenna Test Range Reflectors Employing Width and Height Modulated Triangular Serrations', *Journal of the Institution of Engineers (India)*, **ET, 84**, July 2003.

5. P. Siddaiah and P.V. Subbaiah, 'Viability of On-Off Triangular and On-Off Rectangular Serrations for Improved Performance of Compact Antenna Test Ranges', Accepted for publication of the forthcoming AMSE Journal, France.

6. P. Siddaiah and P.V. Subbaiah, 'Combination of Convex and Concave Serrations for Improved Performance of Compact Antenna Test Range Reflectors', NCSSS-2002, PSG College of Technology, Coimbattore, 1-2 March, 2002.

7. P. Siddaiah and P.V. Subbaiah, 'Performance Augmentation of CATRs Using Width Modulated Exponential And Width and Height Modulated Triangular Serrations', MMTA-2002, School of Environmental Sciences, JNU, New Delhi, 4–6 February, 2002.

8. P. Siddaiah and P.V. Subbaiah, 'Analysis of Width and Height Modulated Exponential Serrated Compact Antenna Test Range Reflectors', Recent Trends in Communication Technology, Jiwaji University, Gwalior, 13-14 April, 2002.

9. P. Siddaiah and P.V. Subbaiah 'Performance Evaluation of WHME and WME Serrated CATR', Recent Trends in Communication Technology, Jiwaji University, Gwalior, 13-14 April, 2002.

10. P. Siddaiah, and P.V. Subbaiah, 'A Comparison Between WMET and WHMET Serrated CATRs', *IETE Journal of Research*, **49**, No. 4, July-August 2003.

Topics in Electromagnetic Waves: Devices, Effects and Applications
Edited by J. Behari
Copyright © 2005, Anamaya Publishers, New Delhi, India

5. Radiation Properties of Right Triangular Microstrip Antenna

V.K. Tiwari, D. Bhatnagar, J.S. Saini, V.K. Saxena and P. Kumar*

Microwave Laboratory, Department of Physics, University of Rajasthan,
Jaipur-302004, India

*Communication System Group, ISRO Satellite Center, Bangalore-560017, India

Abstract: A general right triangular microstrip antenna (RTMA) with angles $(\phi, 90, 90 - \phi)$ has been analyzed in this article. On giving different values to ϕ, radiation properties of two right triangular microstrip antenna, i.e. (45-90-45) and (60-90-30) are theoretically investigated and obtained results are compared with simulation results. The theoretical analysis is carried out by applying cavity model based modal expansion technique while simulation results are obtained with IE3D simulation software. A nice agreement validates the proposed technique.

Introduction

The compact size, lightweight microstrip antennas are becoming increasingly popular for hand held mobile systems, DTH, GPS and DBS systems [1] due to their inherent properties like conformal nature, ruggedness and low fabrication cost. Out of different geometries investigated in recent years, rectangular and circular geometries are the most popular geometries [2]. Since a triangular structure occupies less area on the host surface in comparison to a rectangular patch antenna, efforts to investigate equilateral triangular microstrip patch antennas started in recent times [3-4]. In this article, a general right triangular microstrip antenna with angles $(\theta, 90, 90 - \theta)$ has been analysed and its radiation parameters in two cases, i.e. RTMA geometries with angles (45-90-45) and (60-90-30) are compared with simulation results.

Far Field Radiation Patterns

The geometry and co-ordinate system of a right-angled triangular patch antenna are shown in Fig. 1. The microstrip patch having sides a, b and $\sqrt{a^2 + b^2}$ is lying in XY plane over a dielectric substrate of thickness h, relative permittivity ε_r and loss tangent $\tan \delta$. For the right triangle, $b = a \tan \theta$. The ground plane is considered much larger than the patch dimensions. The theoretical analysis of the proposed probe fed patch geometry is carried out by applying cavity model based modal expansion technique because it requires less computational efforts but provides a good compromise between complexity of modes and accuracy of results. The region between the patch and the ground plane is treated as a cavity bounded by magnetic walls along the edges of the patch and electric walls above and below.

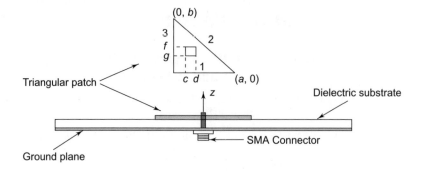

Fig. 1   Geometry and coordinate system for a RTMA structure.

The electric fields components within the substrate are considered z-directed and total electric field at the aperture of antenna are written as the sum of the fields associated with different modes, i.e.

$$E_z = E_0 \left[\cos\left(\frac{m\pi x}{a}\right)\cos\left(\frac{n\pi y}{b}\right) + (-1)^{m+n}\cos\left(\frac{n\pi x}{a}\right)\cos\left(\frac{m\pi y}{b}\right) \right] \quad (1)$$

with

$$E_0 = j\omega\mu\sum\sum \frac{J(d-c)(f-g)\left[\cos\left(\frac{m\pi(d-c)}{2a}\right)\cos\left(\frac{n\pi(f-g)}{2b}\right) + (-1)^{m+n}\cos\left(\frac{n\pi\,(d+c)}{2a}\right)\cos\left(\frac{m\pi\,(f+g)}{2b}\right)\right]}{\left(K^2 - K_{nm}^2\right)\frac{ab}{2}} \quad (2)$$

Here the patch is excited in such a way that the input filamentary current at feed location (x, y) is

$$J_z = J c < x < d \ \text{ and } \ g < y < f$$

$$0 \text{ elsewhere} \quad (3)$$

Since the geometry of antenna is considered probe fed, the modeling of coaxial feed is done by assuming a current ribbon of effective thickness $2w$ centered on the feed axis at a distance OD from point O lying on the patch. The resonance frequency of RTMA structure in TM_{nm} mode is given by

$$f_r = \frac{K_{nm} \cdot c}{2\pi\sqrt{\varepsilon_r}} \quad (4)$$

where
$$K_{mn} = \sqrt{\frac{m^2\pi^2}{b^2} + \frac{n^2\pi^2}{a^2}} \qquad (5)$$

The effect of fringe fields is incorporated in the present analysis by replacing side length a and b of the patch by effective side lengths a_e and b_e given as

$$a_e = a + \left(\frac{h}{\sqrt{\varepsilon_r}}\right) 1.488$$

$$b_e = b + \left(\frac{h}{\sqrt{\varepsilon_r}}\right) 1.488 \qquad (6)$$

On applying the concept of equivalent sources and image, the far zone radiation fields E_θ, E_ϕ are determined. These expressions are then applied to determine E plane ($\phi = 0$) and H plane ($\phi = \pi/2$) radiation patterns factor R_{th} ($\phi = \pi$) in TM_{11} mode of excitation given by

$$R_{th}(\theta, \phi) = \left[|E_\theta|^2 + |E_\phi|^2\right] \qquad (7)$$

The E and H plane radiation patterns are drawn in Figs. 2 and 3 for two RTMA geometries having $\theta = 45$ and $30°$, respectively, and are compared with simulation results. The computed and simulated 3 dB beam width of the radiation patterns of both the antennas are in good agreement.

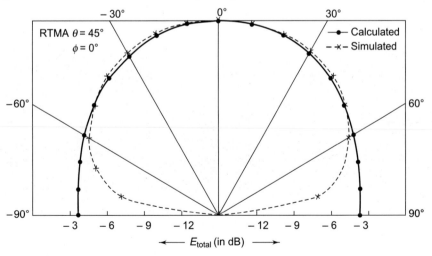

Fig. 2 E-plane radiation patterns of RTMA geometry with $\theta = 45°$.

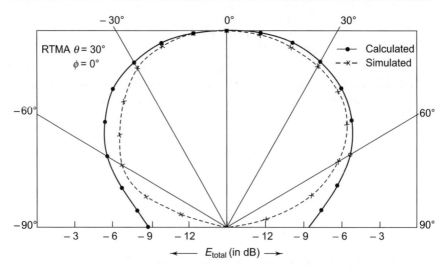

Fig. 3   *E*-plane radiation patterns of RTMA geometry with θ = 30°.

Other Antenna Parameters

An efficient use of microstrip elements requires the knowledge of both field configuration and the impedance seen by the feed. Proper feed point is obtained by varying the feed point on the patch so that complex antenna impedance matches with the impedance of feed line. The expressions for complex antenna impedance Z_{in} in TM_{nm} mode of excitation is obtained in terms of voltage at the feed following [5] and is given as

$$Z_{in} = \frac{j\omega\mu_0}{a_e b_e}$$

$$\sum_{m=1}^{4}\sum_{n=1}^{4}\left[\left[\frac{\left[\left(\cos\left(\frac{m\pi x_0}{a_e}\right)\cos\left(\frac{n\pi y_0}{b_e}\right)+(-1)^{m+n}\cos\left(\frac{n\pi x_0}{a_e}\right)\cos\left(\frac{m\pi y_0}{b_e}\right)\right)\right]^2}{K^2\left(1-j\delta_{eff}\right)-K_{mn}^{2}}\right]\right]$$

(8)

To include the effect of radiation losses, conductor losses and dielectric losses on input impedance, propagation constant K is replaced by K_{eff} which contains $\tan\delta_{eff}$ in place of $\tan\delta$ where

$$\tan\delta_{eff} = 1/Q_t \qquad (9)$$

The total quality factor Q_t includes radiation losses, dielectric losses and conductor losses. The variations of computed and simulated input impedance and directivity of RTMA geometry with θ = 45° are shown in Figs. 4 and 5, respectively. A fairly nice agreement between computed and simulated results validates the proposed technique of treating RTMA geometries.

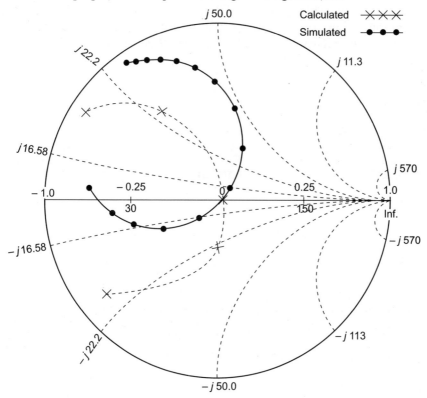

Fig. 4 Variation of computed and simulated input impedance with frequency (θ = 45°).

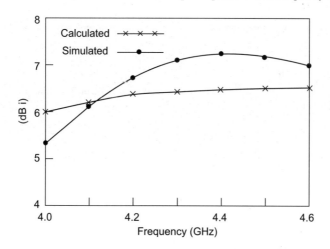

Fig. 5 Variation of computed and simulated directivities with frequency (θ = 45°).

In Table 1, the computed parameters of two RTMA geometries are compared among themselves. The overall performance of RTMA geometry with $\theta = 45°$ is better than that with $\theta = 30°$. The work on other RTMA geometries is in progress. The comparison of the performance of RTMA geometries with similar rectangular microstrip antenna indicate that the performance of these antennas are comparable to that of rectangular microstrip antenna, hence, these antennas may prove to be a better choice in array applications.

Table 1.   Summary of computed results of RTMA geometries at resonance frequency

Parameters	RTMA geometry ($\theta = 45°$)	RTMA geometry ($\theta = 30°$)
Patch dimensions	$a = 30$ mm, $b = 30$ mm	$a = 30$ mm, $b = 17.32$ mm
Resonance frequency	4.224 GHz	5.955 GHz
Bandwidth	0.7215%	0.533%
Directivity	6.107 dB	6.38 dB
Radiation efficiency	84.96%	81.01%
R factor	98.004	132.58

$\varepsilon_r = 2.55$, tan $\delta = 0.0009$, $\sigma = 5.8 \times 10^7$ S/m, $h = 0.16$ cm.

Acknowledgement

Authors are thankful to Dr. S. Pal and Dr. V.K. Lakshmisha, ISRO Bangalore, for permitting them to use simulation facilities at their center.

References

1. Garg, R., Bhartia, P., Bahl, I. and Ittipiboon, A. Microstrip antenna design hand book, Artech House, Norwood, 2001.

2. Wong, K.L. Compact and broad band microstrip antennas, John Wiley & Sons, Inc, 2002.

3. Kumprasert, N. and Kiranon, W. *IEEE Trans Antennas & Propag (USA)*, **42** (1994) 178.

4. Lu, J.H., Tang, C.L. and Wong, K.L., *IEEE Trans Antennas & Propag (USA)*, **47** (1999) 1174.

5. James, J.R. and Hall, P.S. Handbook of microstrip antenna, Peter Peregrines Ltd., London, 1989.

Topics in Electromagnetic Waves: Devices, Effects and Applications
Edited by J. Behari

6. Mobile Phone Antenna and Human Body Interaction

J. Behari

School of Environmental Sciences, Jawaharlal Nehru University,
New Delhi-110067, India

Abstract: The recent developments in mobile communication has drawn urgent attention to the biological effects of electromagnetic fields. Hand held mobile phones are generally used as cordless phones in various positions with respect to the body. There are two distinct possibilities by which health could be affected as a result of RF field exposure. These are thermal effects caused by holding mobile phones close to the body. Secondly, there could be possible nonthermal effects from both phones and base stations. Some people may be adversely affected by the environmental impact of mobile phone base stations sited near their homes, schools or any other place. Apart from the controversies over the possible health effects due to the non-thermal effect of EMF's, the electromagnetic interaction of portable radio with human head need to be quantitatively evaluated. While a lot of efforts have gone into the latter, a clear picture has yet to emerge. Recent advances and the problems are presently discussed.

Introduction

With the ever increasing use of the cellular telephones and the personal communication services, special attention has been drawn to the biological effects of electromagnetic fields, which has remained a subject of continuing concern. Wireless communication systems operate at several frequencies in the electromagnetic spectrum. In the USA it operates at two frequencies, viz. the old existing ones at 850 MHz and the newer personal communication services at 1900 MHz. In terms of the electromagnetic spectrum, cell phones fall between microwave ovens and TV transmitters. Such radiation, though non-ionizing, can induce biologically significant heating (Fig. 1). European mobile phones operate at slightly different frequencies than the above. The energy corresponding to these frequencies is insufficient to knock an electron from atoms in a living tissue and belong to the nonionizing part of the electromagnetic spectrum. A commonly occurring partial body exposure of humans to microwave radiation occurs with the use of cellular phones. Because the phone antenna is close to the head, much effort has gone into determining the dosimetry profile of microwaves in the head in various possible configurations. The geometry of holding the mobile phones suggest that the exposure will be principally to the side of the head for the hand held use, or to the other parts of the body closest to the phone during hand-free use. Frey (1998) made the argument that the headaches were linked to microwave emissions from cellular phones. The body of research is controversial in several aspects since the experimental results are mostly understood in terms of thermal effects. However, the effects due to nonthermal effects are controversial and in a way are poorly understood. The accepted existence of nonthermal phenomena points to biological organization beyond the realm of known laws of physics and may be beyond limits set by chemical reactions in biomolecular systems.

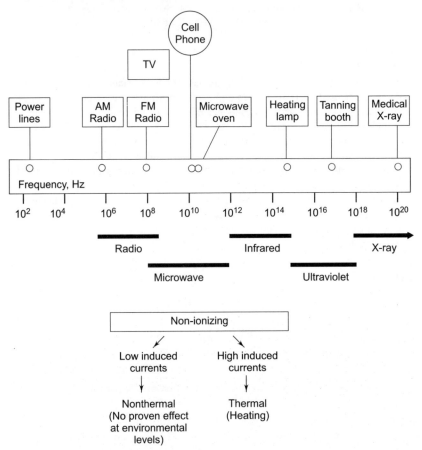

Fig. 1 Position of cell frequencies phones between microwave ovens and TV transmitters.

The animal research that has subsequently followed provides possible importance of localized exposures and the occurrence of hot spots on biological objects. Development of brain cancer takes years or may be decades to develop and in view of the difficulty in controlling the exposure parameters over such a long time it is difficult to reach any definite conclusion. Exposing rats to pulse modulated 837 MHz RF energy similar to that emitted by some digital cell phones does not cause any promotion of brain cancer (Adey et al. 1999). Adey et al. reported the same finding for continuous wave radio frequency (RF), such as those emitted by analog cell phones.

Physics of the Problem

In pursuance of the above, considerable efforts have gone in understanding the complexity of mode of interaction of the antenna of hand held phone and the human body (head). The most commonly used are hand held mobile phones. Mobile telecommunication equipment consists of a single unit with integrated microphone, ear piece, transreceiver and antenna. Mobile phones operate at low power levels: 600 mW in case of analog, while 125 mW in case of digital phones. Since these are placed close to the head, therefore, it is speculated that these may cause exposure level limit close or above the prescribed limit. Fields induced inside the head depends upon the position of the handset with respect to the head and the electric properties of the tissues. The problem is further compounded because different types of antennas cause different amount of field induced in the human head (Jansen and Rahmat-Samii 1995). The telephone antennas are modeled as a dipole or monopole on a metal box (Fig. 2). These have led to (a) characterization of the performance of antenna of the cellular phones and to improve upon it at various distances of the antenna with respect to the head in various configuration and (b) to compute the RF energy deposition in various tissues and to calculate the specific absorption rate (SAR). This then can be compared with the suggested standards (Jansen and Rahmat-Samii 1995, Stuchuly 1994, US ANSI Standard 1992).

Basic to the understanding of the implications of the mobile phones is the knowledge of the induced fields. A reduction in the output of the mobile phone would require the need to have closely spaced base stations, which may not go well with the environmentalists. Moving antennas to a larger distance from the user's head would require a larger handset size, again contrary to the requirement of user. The computation of the induced fields in head is complicated and depends upon a number of parameters, viz. (i) operational frequency and the antenna input power, (ii) position of the device with respect to the head, (iii) design of the device vis-á-vis the outer shape of the head and (iv) the dielectric properties of the tissue and the shape of the head. Out of these (iii) and (iv) are the subject of the technical details and the life history of the individual. Subsequent to the above computation of SAR in the human head due to the various approaches using anatomically derived structure, based on magnetic resonance imaging were adopted for homogenous and heterogeneous head shaped volume (Jansen and Rahmat-Samii 1995, Dimbylow and Mann 1994, Gandhi 1995, Martens et al 1995). Most of these authors have used finite difference time domain technique (FDTD) for the analysis because of the flexibility and efficiency in solving complex heterogeneous geometries. Okoniewski and Stuchly (1996) used Yee cell rectangular grid (Yee 1996) and the total field formulation (Taflove 1995). The computational space was truncated by a perfectly matched layer (Berengner

1994) of 7 cell thickness with a parabolic profile to ensure reflections below at least 40 dB. Two mesh sizes used were either 5 mm or 3.4 mm. In cases where the metal surfaces did not coincide with the mesh, an algorithm was used that allows for accurate treatment of fields near these surfaces (Anderson et al. 1996). These authors obtained far field antenna pattern from the computed near field electric and magnetic field vectors on a box enclosing the modeled structures using the field equivalence principle (Taflove 1995). The antenna efficiency is calculated as

$$\eta = P_{rad}/(P_{rad}+P_{abs})$$ (1)

where P_{rad} is the power radiated by the antenna and P_{abs} the power absorbed by the head. These authors used the distance d as that between the monopole and the closest head surface. The following formulation applies: (1) the steady state radiated power from the antenna is 1 W (i.e. the power radiated in the far field plus the power dissipated in the head and hand) and (2) the antenna patterns are normalized to those obtained for the Yale head model and for 3 cm separation between the handset and the head using 5 mm FDTD mesh.

The box model of the head at close distance to the phone give lower antenna efficiency and higher absorbed power in the head, and as such higher SARs in comparison to other models. Spherical models provide estimates of the antenna efficiency and total absorbed power that are in reasonably good agreement with relatively low resolution head models. These authors have further concluded that spherical (and box) models of the head give overestimated values of the decrease in antenna efficiency as well as of the total power absorbed and various local and limited volume averaged SARs. The box representing the head nearly entirely blocks the radiation in the head direction. In the case of sphere there is a significant reduction in power radiated towards the half space where the head is located.

Okoniewski and Stuchly (1996) have described three type of models, viz. dielectric covered metal box for a three-layered tissue box, sphere and a anatomically correct (Yale) head model. The computations were carried out at 915 MHz. They concluded that the effect of the anatomically correct head is well stimulated (±3 dB) by a three layered sphere. This result further confirms the findings of Jansen and Rahmat-Samii (1995) who used a less anatomically correct model of the head and a different antenna. The differences they computed were somewhat bigger, which are explained by the different antenna geometry. As the distance increases the antenna pattern is less affected and less power is radiated toward the head direction.

Effect of the Hand Holding a Hand Set

The mode of holding the phone is shown in Fig. 2. It is apparent that the hand is capable of absorbing a significant amount of the antenna output power and thereby, affecting the performance of the telephone. Correspondingly, its presence decreases SARs deposition in the head. A reduction of the power absorbed by the hand by over 50% can be accomplished by the use of a choke. Some of the input parameters to these are the tissues constituting the head model comprises muscle, bone, skin, brain, fat, eyeball and lens. The mass density of each tissue is taken to be 1000 kg/m^3.

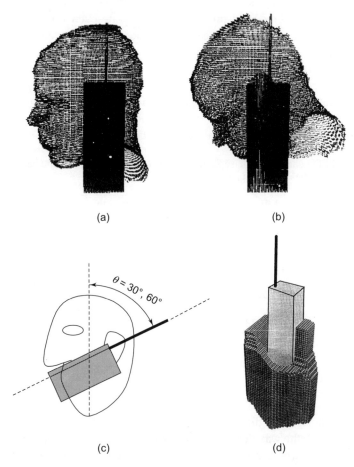

(a)   (b)

(c)   (d)

Fig. 2 The two head models with the telephone used for the calculations : (a) vertical, (b) tilted 30° relative to vertical, (c) portable radio models and their position relative to the head and (d) model of the hand holding the radio.

Generally, the hand holding the handset is made up of a metal covered with a thin lossless dielectric. This decreases the radiated power due to the power absorption in the hand. A simple unoptimized choke that provides an apparent open circuit on the upper surface of the antenna box improves the antenna efficiency from 61 to 70% compared to 77 % without the hand (separation of 3 cm between the antenna and the head). Correspondingly, the peak SAR in the head decreases to 5.8 W/kg from 14 W/kg.

Antenna Head Separation

Effects of the changes in the separation between the antenna and the head model (915 MHz, 1 W) is shown in Table 1. Changes in the antenna efficiency, power deposited and SARs with separation between the antenna and various models of the head suggest that the box head model is not adequate even for larger separations. The similarities, in terms of the result obtained between the spherical model and the simpler head model without an ear remains even for greater separations. A comparison between the Gent and Yale heads suggest that the 3.0 and 1.5 cm are equivalent. This observation is further supported by an approximate separation due to the ear (thickness). These results are in good agreement with those reported elsewhere (Jansen and Rahmat-Samii 1995) and SAR in 1 g of tissue (Jansen and Rahmat-Samii 1995, Douglas et al. 1995).

Table 1.   Effects of the change in the separation between the antenna and the head model; 915 MHz, 1 W (Okoniewski and Stuchly 1996)

Model	Separation	Efficiency	P_{abs} (W)	SAR in the head (W/kg)		
				Peak	1 g	10 g
Homogeneous box	1.5	16	0.84	18.1	14.1	9.25
	3.0	51	0.49	3.2	2.6	1.9
Homogeneous sphere	1.5	46	0.54	13.4	10.9	7.0
	3.0	73	0.27	3.7	3.0	2.1
Gent head	1.5	51	0.49	11.2	8.6	4.8
	3.0	72	0.28	3.2	2.4	1.4
Yale head	1.5	60	0.40	3.5	2.65	1.8
	3.0	77	0.23	1.7	1.1	0.8

The hand held portable radios are operated at 900 MHz or at 1500-1800 MHz, and as such the wavelengths of the EMF's which they produce are much shorter than the whole human body. It is assumed in almost all the computations that when the radio is located within the vicinity of the head, the body below does not affect the SAR distribution within the head. Watanabe et al. (1995) examined the validity of the SAR distribution in an isolated head model and compared with that in a whole body model when both models were exposed to the near field of a short electric dipole at 300 MHz. Watanabe et al. (1996) have also carried out SAR distribution in $1 \times 1 \times 1$ cm cubical cells. The calculations were carried out using FTDT method. These authors reported that the distribution on the surface of the isolated head model agreed well with that of the whole body model: the difference of the maximum local SARs on the surface was within 2%. The difference between the local SARs calculated with these models were as much as 37% in the deep region of the head, the absorption in the deep region is much smaller than the surface absorption and is not significant when compared with it. The hand held portable radio was assumed to be a $135 \times 42.5 \times 25$ mm conducting box with either ½-wavelength dipole antenna or a ¼-wavelength monopole antenna on its upper plane (Fig. 3). It was located on the left side of the head. These authors rotated hand and a radio by 60° so that the radio model was positioned between the operator's mouth and ear. They reported that the radiation pattern of the antenna is significantly modified with the user's hand.

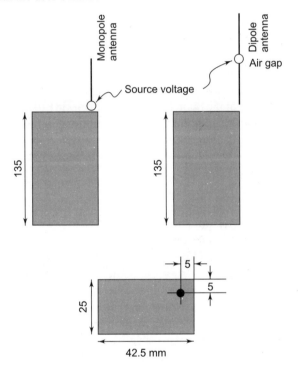

Fig. 3   Portable radio models and their positions relative to the head.

It was concluded that:

1. The maximum local SAR decreases roughly in proportion to $d^{-3/2}$. This is independent of whether the antenna is a ½λ dipole or a ¼λ monopole. This also does not depend on the frequency (900 MHz or 1.5 GHz).
2. The maximum local SAR decreases with increasing antenna length (¼λ to ½λ).
3. When the auricle is located in the vicinity of the head, the maximum local SARs in the homogenous head model agree well with those in the heterogeneous model.
4. The maximum local SARs in the head do not depend on the position of the head holding the radio as long as the antenna is not obstructed by the hand or by other parts of the body.

These authors also reported that the SAR obtained experimentally were significantly less than those calculated numerically.

SAR Distribution in Various Head Phantoms

Hombach et al. (1996) concluded that (1) the region with high absorption values in all head phantoms is small and close to the feed point of the dipole. In most part of the head the EM field is relatively low. (2) Two SAR maxima can be identified in each of the MRI based phantoms, one on the skin's surface and the other on the brain's surface. The SAR induced in the bone tissue is considerably lower. The absolute values of the maxima differ from model to model. (3) Varying SAR distributions in the brain have been identified with the presence of grey and white matter. This may be due to difference in dielectric parmeters of the two different portions of the brain (Ray and Behari 1988). (4) Since the tissues are uniform inside the brain region, the SAR distribution shows a smooth decay.

In simplified head phantom model with only two type of tissues, the resulting SAR distributions show some agreements and differences with respect to that of the anatomically correct phantoms.

These differences can be directly explained by the energy absorption mechanism. According to Kuster and Balzano (1992) the induced SAR is primarily determined by the square of the H field which drops inversely proportional to the square of the distance from the source. In case of transmitters very close to the head, this factor is dominant so that even lossy bone tissue can be approximated by a layer of air of the same thickness in order to assess the SAR in the brain tissue behind this layer. In other words, the distance dependence dominates the field attenuation. Hombach et al. (1996) have concluded that the spatial peak SAR remains unaffected by the size and the shape of the human head for the electromagnetic source at a defined distance from the human head. Compared to other factors like the distance of the source from the head and the design of the devices, the effects caused by the anatomy are minor especially in the case of volume-averaged values. This is due to the strong radial decay of the H field in the vicinity of the source.

The comparison of the results obtained from the inhomogeneous and homogenous phantoms suggest that the latter are highly suited to be used in compliance tests for hand held mobile telecommunication equipment operating in the 900 MHz band. The overestimation of the averaged spatial peak SAR values is small when compared to the largest value by the inhomogeneous phantoms. This is especially true if the values are averaged over a volume equivalent to 10 g. The major advantage of using simple homogenous phantoms is that the number of tests can be reduced since small shifts of the source parallel to the surface results in almost no changes in the spatial peak SAR values.

In case of inhomogeneous modeling, variations of several dB must be expected for shifts of a few millimeters. Therefore, inhomogeneous phantoms might result in a low SAR values for some positions of the device with respect to the head, which would not represent the actual exposure of the various users. Peak positions is independent of inhomogeneity. Power reflected from the antenna is also neglected by almost all the models, which is anyway not small.

Several studies dealt with a generic model of a cellular telephone at approximately 900 MHz consisting of a monopole on a metal box 15 cm long with a cross section of 5 to 6 cm by 1.3 to 2.5 cm (Jansen and Rahmat-Samii 1995, Toftgard et al. 1993, Dimbylow and Mann 1994, Martens et al. 1995).

Gandhi et al. (1996) examined two different lengths of monopole antenna ($\lambda/4$ and $3\lambda/8$). They used plastic coated handset of dimensions that are typical of newer mobile telephones both at 835 and 1900 MHz. They have also studied the effect of tilting the handset as for typical usage at an angle of 30° relative to vertical position and comparing the two SAR results. By scaling the model of the head and neck to obtain reduced size models representative of 10- and 5-year old children, they calculated the SAR distributions and found deeper penetration of EM energy and SARs for internal tissues that are several times higher than for the model of the adult. Since the tissue properties are not as well characterized, and wide varying values have been reported for fat, bone and cartilage, they have studied the effect that these properties can have on 1 g SARs that need to be examined for compliance with the safety standards. Gandhi et al. (1996) identified a problem with interpreting the safety guidelines (ANSI/ IEEE 1992), since unspecified and different subvolumes in the shape of a cube may be taken in order to obtain peak 1 g SARs that should not exceed 1.6 W/kg for uncontrolled environment. These safety guidelines are given in terms of the maximum permissible exposure of electric and magnetic fields, or of power density for controlled and uncontrolled environment. These exposure limits can be used for far fields but cannot be used for highly nonuniform field such as near fields region of a cellular telephone. An exposure condition can be considered to be acceptable if it can be shown that it produces SARs below 0.08 W/kg, as averaged over the whole body, and spatial peak SAR values not exceeding 1.6 W/kg, as averaged over any 1 g of tissue. These authors used Finite Time Difference Technique to obtain SAR distributions and concluded that longer antennas give SARs that are smaller than those for $\lambda/4$ antennas. They examined the effect of tilting the antenna with respect to head by 30° and found that at 835 MHz the result is a reduction of SAR relative to untilted configuration, since the antenna is now further away from the head for much of its length. However, when the antenna length is small (1900 MHz) the effect is inconsequential. In small models (e.g children) the penetration is higher and SAR values are considerably higher. These authors have also used truncated models (one-half and one-third models) of the head and neck and obtained SARs distributions that are close to those for full models. These authors also studied the effect of using the widely different tissue properties and of using homogenous instead of the anatomically realistic heterogeneous models on the SAR distributions. Homogenous models are shown to grossly overestimate the 1 g SARs values.

Because of the proximity of the hand to telephone, Gandhi et al. (1996) modeled the hand by a region of 2/3 muscle equivalent material of thickness 1.974 cm, covering the handset from the three sides. The remaining side facing the head, with height two-thirds that of the hand set. These authors have reported that homogenous model overestimates 1 g peak SAR by 42% as compared to that obtained using the anatomically based model.

Intracranial Fields Induced by Mobile Phones

Experimental Techniques

The SAR distribution is obtained by determining the electric field with miniaturized E-field probes inside shell phantoms filled with tissue simulating material (Fig. 4). The probes are positioned by a six axis precision robot with a position repeatability of better than ± 0.02 mm at constant temperature (Hombach et al 1996). An optical surface-detecting system is integrated into the probes, which enables the accurate positioning of the probe with respect to the phantom's inner surface. The SAR distribution can be measured for basically any volume. For dosimetric assessments of RF-sources close to the head, the E-field probe scans over a large area inside the head to roughly localize the maximum SAR value. In a subsequent step, SAR measurements are done along a fine grid within a 35 g volume cubically shaped around this maximal value. This cube is large enough to provide enough data to evaluate the spatial peak SAR. The whole procedure is fully automated.

Fig. 4   Human phantom and computer controlled ε field probe

As mentioned above, the most important parameter determining the amount of energy absorbed in the tissues of the head is the distance separating the antenna or radiating structure from the head. Absorption decreases approximately as the inverse square of increasing distance between the antenna or radiating structure and the tissue. At the frequencies of 10-20 GHz, typical attenuation are in the range of 20 dB/mm from the skin surface.

Are Mobile Phones Safe?

Hand held mobile phones are generally used as cordless phones in various positions with respect to the body. There are two distinct possibilities by which health could be affected as a result of RF field exposure. These are thermal effects caused by holding mobile phones close to the body. Secondly, there could be possibly nonthermal effects from both phones and base stations. Some people may be adversely affected by the environmental impact of mobile phone base stations sited near their homes, schools or any other place.

Based on modeling it has been estimated that SAR to head from a 900 MHz cellular telephone vary from 0.16 to 0.69 W/kg, and for the brain 0.06-0.41 W/kg (Gandhi 1995). However, a similar examination by Dimbylow and Mann (1994) with a vertical or a lateral antenna suggested a 3-4 W/kg averaged over 1 g. Excell (1998) calculation suggested higher values upto 4.2 W/kg rising to 8.2 W/kg at 1800 MHz based on magnetic resonance imaging (MRI) and FDTD techniques. A level of 1 W/kg is expected to raise the temperature by less than 0.5° C. Research pertinent to the use of mobile phones by Sarkar et al. (1994) and Lai and Singh (1995, 1996) showed an increase in DNA breaks at 2.45 GHz. More recently, Paulraj and Behari (2004) also obtained similar results at amplitude modulated RF signal (112 MHz-AM 16 Hz). However, it may be mentioned that these findings are not supported by several other workers (Malyapa et al. 1997, Chou et al. 1992) (Table 2). Long term exposure of the mouse (Repacholi et al. 1997) showed an elevated risk of developing lymphoma in a transgenic strain particularly prone to developing the condition. Use of phone in driving stimulator leads to negative reaction time. However, this may be considered a multitasking problem. Evidence for a direct memory effect is on brain slices from the hippocampus of rat that showed changes in long term potentiation when exposed to 915 MHz (Scott et al. 1998). Mild et al. (1998) looking for a subjective response in subjects, suggested increased headache or sensation of warmth when using mobile phones. Braune et al. (1998) reported blood pressure changes induced by exposure to the right side of the head. Preece et al. (1999) studied the effect of a 915 MHz simulated mobile phone signal on

Table 2.   Brain cancer in rats after RF radiation exposure

Researchers	Exposure to RF radiation			No. of rats		
	Frequency	SAR	Duration (days)	RF exposed	Un-exposed	Cancer
	Brain Tumor Generation					
C.K. Chou et al. (1992)	2450 PM	0.15-0.4	25	100	100	None
J.C. Toler et al. (1997)	435 PM	0.32	21	200	200	No significant difference between groups
M.R. Frei et al. (1998)	2450 FM	0.3	18	100	100	None
Paulraj and Behari (2004)	112 MHz-AM 16 Hz		35	10	10	Indications of tumor promotion

cognitive function in man and reported evidence of increased responsiveness, strongly in the analogue and less in the digital simulation, in reaction time. They further concluded that this could be associated with mild localized heating, or possibly a nonthermal response, which is power-dependent.

Radon et al. (2001) showed that pulsed RF electromagnetic fields (900 MHz carrier frequency pulsed with 217 Hz) similar to those emitted from mobile radio telephones had no short term or medium term effects on salivary melatonin, cortisol, neopterin and sIgA concentrations. These authors also confirmed the observation that nocturnal melatonin levels are not affected by exposure to RF electromagnetic fields as also evidenced by Mann et al. (1998). The findings are in confirmation with the data showing that day time melatonin levels are unaffected by exposure to RF electromagnetic fields of 900 and 1800 MHz (de Seze et al. 1999). Also no effect was noted on melatonin synthesis and excretion in humans exposed to 50 Hz magnetic fields of 10 μT (Selmaoui et al. 1996). Vollrath et al (1997) have reported that day as well as night melatonin levels were unaffected. Freude et al. (1998) have concluded that radiation intensity of an usual global system for mobile communication telephone call may alter preparatory bioelectrical activity. However, it may be mentioned that not much data are available in this field of research.

The epidemiological results so far reported are at times contradictory. The animal research does not show any confirmatory trend. The epidemiological results may lack sensitivity for the data collected over a long period of time, may not be a pointer to the development of carcinogenesis due to electromagnetic field exposure alone and the relevance of the animal data to human health is uncertain. However, mobile phone manufacturers' must prove that their product meets exposure limit. The safe exposure limit from mobile phones are largely developed on the basis of whole body exposure and the engineering design of the handset antennas.

There are some general principles. The RF energy from digital handset is usually lower than that from analogues and the RF energy from PCS handsets operating near 1900 MHz reaches more superficial layers of the head than that from cellular handsets operating at 850 MHz.

Exposure Limit

1. The Federal Communication Commission (FCC) has set a limit for the exposure of a user to RF radiation from a wireless hand set which is 1.6 W/kg of body tissue averaged over any gram of tissue. However, its relation to the threshold for real hazard is not clear. The prescribed limit of 1.6W/kg is based on thermal effects of total heat load in animals being subjected to whole body exposure, the averaging requirements based on engineering considerations.

2. There has been general agreement that behavioral disruption is the potentially adverse effect found in animals at the lowest exposure level. The term 'behavioral disruption' refers to the tendency of animals to stop performing a complex learned task when exposed to a sufficient amount of RF energy. The

disruption occurs at a threshold role of SAR of about 4 W/kg of body weight. Taking into account a safety factor of 10, maximum permissible exposure for humans of 0.4 W/kg of body weight. The standard for partial body weight has been set to 8 W/kg, though the rational for this is not very much clear.

Unfortunately, it is not possible to measure and hence the dependence is on SAR inside the head computer modeling.

The hand free devices that move the handset away from the user's body tend to reduce exposure, though these may cause exposure to sperm and there is fear of infertility. The present day controversies regarding the mobile phone is probably because of nonthermal effects.

It is unlikely that low powered transmitters line mobile phones will present thermal hazard. It is possible that there are currently unrecognized health effects from the use of mobile phones. Children are more vulnerable because of their developing nervous system, greater absorption of energy and a longer life time of exposure. As such unnecessary use of mobile phones by children should be avoided. Other users should use the device with the instructions as supplied by the manufacturers.

References

1. Adey, W.R., Byus, C.V., Cain, C.D., Higgins, R.J., Jones, R.A., Kean, C.J., Kuster, N., MacMurray, N.A., Stagg, R.B., Zimmerman, G.J., Phillips, J.L. and Haggren, W. (1999). Spontaneous and Nitrosourea induced primar;y tumors of the central nervous system in fischer 344 rats chronically exposed to 836 MHz modulated microwaves. *Radiation Res.* **152,** 293–302.

2. Anderson, J., Okoniewski, M. and Stuchly, S.S. (1996). Practical 3-D contour/staircase treatment of metals in FDTD. *IEEE Microwave Guided Wave Lett.* **6** (3): 146–148.

3. ANSI, C95.1 (1992). American National Standard Safety levels with respect to human exposure to radio frequency electromagnetic fields, 300 kHz to 100 GHz. New York: IEEE, 1991.

4. Berenger, J.P. (1994). A perfectly matched layer for the absorption of electromagnetic waves. *J Comp. Phys.* **14**: 185–200.

5. Braune, S., Wrocklage, C., Raczek, J., Gailus, T. and Lucking, C.H. (1998). Resisting blood pressure during exposure to a radio frequeny electromagnetc field. *Lancet* **351**: 1857–1858.

6. Chou, C.K., Guy, A.W., Kunz, L.L., Johnson, R. B., Crowley, J.J. and Krupp, J.H. (1992). Long- term, low level microwave irradiation of rats. *Bioelectromagnetics* **13**: 469-496.

7. de Seze, R., Ayoub, J., Peray, P., Miro, L. and Touitou, Y. (1999). Evelutions in humans of the effects of radiocellular telephones on the circadian patterns of melatonin secretion, a chronobiological rhythm marker. *J Pineal Res* **27**:237-242.

8. Dimbylow, P.J. and Mann, S.M. (1994). SAR calculations in an anatomically realistic model of the head for the mobile communication transreceivers at 900 MHz and 1.8 GHz. *Phys Med Biol* **39**:1537–1553.

9. Douglas, M.A., Okoniewski, M. and Stuchly, M.A. (1995). Modeling of wireless personal communication transmitters transmitters in the presence of the user's

body. Wireless' 95-7th Int Conf. Wireless Communicat., Calgary, Alberta, pp. 77–83.

10. Excell, P. (1998).Computer modeling of high frequency electromagnetic field penetration into the human head. *Measurement and Control* **31**: 170–175.

11. Frei, M.R., Jauchem, J.R., Dusch, S.J., Merritt, J.H., Berger, R.E., Stedham, M.A. (1998). Chronic, low-level (1.0 W/kg) exposure of mice prone to mammary cancer to 2450 MHz microwaves. *Radiat Res* **150** (5): 568-576.

12. Freude, G., Ullsperger, P., Eggert, S. and Ruppe, I. (1998). Effects of microwave emitted by cellular phones on human slow brain potentials. *Bioelectromag.* **19**:384–387.

13. Frey, A.H. (1998). Headaches from cellular telephones: Are they real and what are the implications? *Environ Health Prospect* **106**, 101–103.

14. Gandhi, O.P. (1995). Some numerical methods for dosimetry: Extremely low frequencies to microwave frequencies. *Radio Sci* **30**: 161-177.

15. Gandhi, O.P., Lazzi, G. and Furse, C.M. (1996). Electromagnetic absorption in the human head and neck for the mobile telephones at 835 and 1900 MHz. *IEEE Trans MTT* **44:** 1884–1897.

16. Hombach, V., Klaus, M., Burkhardt, M., Kuhn, E. and Kuster, N. (1996). The dependence of EM energy absorption upon human head modeling at 900 MHz. *IEEE MTT 44* (10) 1865-1873.

17. Jansen, M.A. and Rahmat-Samii, Y. (1995). EM interaction of hand set antennas and a human in personal communications. Proc. IEEE **83**: 7-17.

18. Kuster, N. and Balzano, Q. (1992). Energy absorption mechanism by biological bodies in the near field of dipole antennas above 300 MHz. *IEEE Trans Veh. Technol.* **41**(1): 17–23.

19. Lai, H. and Singh, N.P. (1995). Acute low intensity microwave exposure increases DNA single-strand breaks in rats brain cells. *Bioelectromagnetics* **16**: 207–210.

20. Lai, H. and Singh, N.P. (1996). Single and double –strands DNA breaks in rat brain cells after acute exposure to radiofrequency electromagnetic radiation. Internatl. *J Radiat. Biol* **69** : 513–521.

21. Malyapa, R.S., Ahern, E.W., Straube, W.L., Moros, E.G., Pickard, W.F. and ROTI ROTI, J.L.l. (1997).Measurement of DNA damage by alkaline comet assay in rat brain cells after in vivo exposure to 2450 MHz electromagnetic radiation. In proceedings of second world congress for electricity and magnetism in Biology and Medicine, Bologna, Italy.

22. Mann, K., Wagner, P., Brunn, G., Hassan, F., Hiemke, C., Roschke, J. (1998). Effects of pulsed high frequency electromagnetic fields on the neuroendocrine system. *Neuroendocrinology* **67**: 139–144.

23. Martens, L., De Moerloose, J. and De Zutter, D. (1995). Calculation of the electromagnetic fields induced in the head of an operator of a cordless telephone. *Radio Sci* **30:** 283–290.

24. Mild, K.H., Oftedal, G., Sandstrom, M., Wilen, J., Tynes, T., Haugsdai, B. and Hauger, E. (1998). Comparison of systems experienced by users of analogue and digital mobile phones: a Swedish-Norwegian epidemiological study. Report for the National Institute for working life.

25. Okoniewski, M. and Stuchly, M.A. (1996). A study of the handset antenna and human body interaction. *IEEE Transactions on Microwave Theory and Techniques,* **44,** 1855–1864.

26. Paulraj, R. and Behari, J. (2004). Radio frequency radiation effects on protein kinase C activity in rats' brain. *Mutation Research* **545,** 127–130.

27. Preece, A.W., IWI, G., Davies-Smith, A., Wesnes, K., Butler, S., Lim, E. and Varey, A. (1999). Effect of 915 MHz simulated mobile phone signal on cognitive function in man. *Int. J. Radiat. Biol.* **75**(4): 447-456.

28. Radon, K., Parera, D., Rose, D.M. and Vollrath, L. (2001). No effects of pulsed electromagnetic fields on Melatonin, Cortisol and selected markers of the immune system in man. *Bioelectromagnetics* **22:** 280-287.

29. Ray, S. and Behari, J. (1988). Dielectric dispersion in rat brain tissue. In: Microwave technology and applications, edited by Sitaram, R.V.S, Srivastava, G.P. and Nair, P.G. Proceedings of the First Asia-Pacific Microwave Conference, 1986, New Delhi, India.

30. Repacholi, M.H., Bosten, A., Gebski, V., Noonan, D., Finnie, J. and Harris, A.W. (1997). Lymphomas in Em-Pim 1 transgenic mice exposed to pulsed 900 MHz electromagnetic fields. *Radiation Res.* **147:** 631-640.

31. Sarkar, S., Ali, S and Behari, J. (1994). Effect of low power microwave on the mouse genome: A direct DNA analysis. *Mutation Research* **320,** 141-147.

32. Scott, I.R., Wood, S.J. and Tattersall, J.E.H. (1998). Effects of radiofrequency radiation on long term potentiation in rat hippocampus slices. In proceedings of Bioelectromagnetics Society 20th Annual Meeting , St Petersburg, Florida USA.

33. Selmaoui, B., Lambrozo, J., Touitou, Y. (1996). Magnetic fields and pineal function in humans. Evaluation of human function .Evaluation of nocturnal acute exposure to extremely low frequency magnetic fields on serum melatonin and urinary 6-sulfatoxymelatonin circadium rhythms. *Life Sci* **58:**1539–1549.

34. Stuchuly, M.A. (1994). Wireless Communications and the safety of the user. Intl. *J.Wireless Information* **1:** 223–228.

35. Taflove, A. (1995). Computational electrodynamics: the finite-difference time domain method. Artech House.

36. Toftgard, J., Hornsleth, S. and Bach Andersen, J. (1993). Effects on portable antennas of the presence of a person. *IEEE Trans Antennas Propagat.* **42:**739–746.

37. Toler, J.C., Shelton, W.W., Frei, M.R., Merritt, H.R. and. Stedham, M.A. (1997). Long-Term, Low-Level Exposure of Mice Prone to Mammary Tumors to 435 MHz Radiofrequency Radiation. *Radiat Res.* **148:** 227–234.

38. Vollrath, L., Spessert, R., Kartzsch, T., Keiner, M., Hollmann, H. (1997). No short term effects of high frequency electromagnetic fields on the mammalian pineal gland. *Bioelectromagnetics* **18:**376–387.

39. Watanabe, S., Taki, M., Nojima, T. and Fujiwara, O. (1996). Characteristics of the SAR distributions in a head exposed to electromagnetic fields radiated by a hand-held portable radio. *IEEE MTT* **44** (10): 1874–1883.

40. Watanabe, S., Tanaka, T. and Taki, M. (1995). Effects of the body on the SAR distribution in the head of a human model exposed to the near field of a small radiation source and to the far field of the source. In Proc. Convention Rec. IEICE Japan, IEICE, p. B-241.

41. Yee, K.S. (1996). Numerical Solution of initial boundary value problems involving Maxwells equations in isotropic media. *IEEE Trans Antenna Propagat.* **14:** 302-307.

Topics in Electromagnetic Waves: Devices, Effects and Applications
Edited by J. Behari
Copyright © 2005, Anamaya Publishers, New Delhi, India

7. Effect of Radiofrequency Radiation on Biological System

S. Sarkar, K. Chandra, R.C. Sawhney and P.K. Banerjee

Defence Institute of Physiology and Allied Sciences, Lucknow Road,
Delhi-110054, India

Abstract: Biological effects with possible long-term health consequences during and after exposure to electromagnetic fields have been a matter of public concern due to increasing opportunities for exposure to such fields in modern life. Scientific research into possible health effects has been unable to keep pace with the rapid advances in the applications of RF fields in our working and living environment. This delay has led to widespread concern that there are unresolved health issues that need to be addressed as a matter of urgency. In studies with exposure to radiofrequency electromagnetic radiation, likelihood of effects from such exposures indicate reproductive effects, changes in blood counts, increased somatic mutation rate in white blood cells and increased incidence of childhood, testicular and other cancers. In addition, generalized increased disability rates, lenticular opacity, sensitivity reactions have also been reported. Many findings also point towards potential carcinogenic and other health effects of radiofrequency radiation. It is suggested that 'prudent avoidance' of unnecessary exposure should be the order of the day.

Introduction

Radiofrequency (RF) radiation is a subset of electromagnetic energy covering the frequency range 3 kHz to 300 GHz; microwaves (MW) and millimeterwaves (MMW) (300 MHz to 300 GHz) are a subset of the RF spectrum (Fig. 1). India relies enormously on the use of electromagnetic radiofrequency energy for its various projects both in the defence and the civilian sector. Major development projects include radar surveillance, defence communication links and terminals, high voltage power lines, large telecommunication facilities, commercial satellite communication, television broadcasting, geosynchronous satellite ground stations, microwave terrestrial link system etc. A number of microwave components and antennae have been designed and developed for use in digital signal processing techniques for communication, collection and analysis of intelligence signals from communication radar and other non-communication emitters etc. Expertise in microwave applications also include wide bank search, medical linear accelerators, atmospheric instrumentation, marine navigational aids and industrial applications. The nature of media used in telecommunication facilities range from the optical fiber based links to microwave digital radio links, infrared links and also low frequency wireless links. Cellular phones entered the country around 1994 and today there are approximately 33.3 million cell phone users. The Indian Space program utilizes microwave satellite communication, satellite remote sensing for resource survey and management, environmental monitoring and meteorological services including disaster warning. New applications like digital audio broadcasting, internet services, distance education, compressed digital TV services etc. are underway. There has been an impressive expansion of television which now provides access to over 80 per cent of the country's population with more than thousand TV transmitters and direct reception sets. Ranking second in population, the country probably has the highest density of population in the cities and industrial belts with power distribution nets matching those of large developed nations and it also competes with most advanced countries in heavy electrical industries. New technologies using RF frequencies are on rise: millimeter wave frequencies are being exploited for electronic warfare, development of exotic weapon systems like directed weapon systems (DEW), kinetic weapon system (KEW), ray guns for crowd control, remote viewing, psychotropics. Thus, it is probable that apart from service personnel and occupational workers, a large quantum of civilian population is exposed to radiopropagation environment and consequently to the associated hazards of electromagnetic radiation, although, a census figure on these aspects is not yet available.

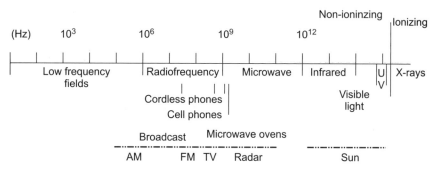

Fig. 1   Electromagnetic spectrum (Source: RSC. EPR 99-1).

Conclusive evidence on the safety or risk of such exposures is not yet established but a growing body of scientific evidences report of bioeffects and adverse health effects which may be possible if not probable (Sarkar et al 1997; Sarkar and Selvamurthy, 2001). The main difficulty in understanding the biological effects of RFR is caused by the complex interaction of the different exposure parameters as well as the mass, shape and size of the body, the orientation of the body with the field vectors, and the electrical properties of both the body and the environment. Other variables which may play a role in possible biological effects are those that characterize the environment (ambient temperature, air velocity, humidity and body insulation) and those that characterize the individual (age, gender, activity level, debilitation and/or disease). Independent variables of such complexity are unprecedented in any other field of biological research. These factors not only make the distribution of energy absorbed in an irradiated organism extremely complex and non-uniform but also lead to the formation of so called 'hot spots' of concentrated energy in the tissue. Knowledge of possible health effects of RFR is still inadequate and inconclusive. Just as studies indicating these fields as being health hazards have been published, so are the studies indicating no risk at all. Health effects are biological changes induced in an organism which may be detrimental to that organism while the biological effects are physiological, biochemical or behavioural changes induced in an organism, tissue or cell (RSC. EPR 99-1).

Frequency Ranges

The effect of exposure to electromagnetic radiofrequency field depends on the frequency of the field. In the lower frequency range of 300 Hz to 1 MHz, currents are induced in the body which may influence the biological system and interfere with normal information processing capabilities of the central nervous system. The relevant dosimetric quantity in this frequency range is current density, expressed in mA/m^2.

In the intermediate frequency range of 100 kHz to 10 GHz, absorption of electromagnetic energy leads to generation of heat and heat tolerance is the limiting factor. The appropriate dosimetric quantity is the specific absorption rate (SAR), i.e. the rate at which energy from electromagnetic field is absorbed per unit mass and is expressed in W/kg. The transition from current density to SAR becoming the relevant dosimetric quantity is very gradual. Thus, in the frequency range of 100 kHz to 10 MHz both quantities apply. With respect to SAR, biological effects are thermal if >1 W/kg or nonthermal if < 0.1 W/kg; between the thermal and the nonthermal effect, there is a range of mechanism of action, assumed but not yet explained.

In the upper frequency range of 10 GHz to 300 GHz, the energy from electromagnetic field is increasingly dissipated at body surface, resulting in heating of superficial layer, mainly the skin. This heating is directly related to the power density of the incident field. Therefore, the power density is considered to be the relevant dosimetric quantity in this frequency range measured in W/m^2.

Basic Restrictions

Biological effects of radiofrequency electromagnetic field have been reviewed by expert committees of various countries like the Netherlands (Summary of Advisory Panel, 1998), the United States of America (IEEE, 1999) and Canada (RSC.EPR 99-1, 1999). Many other countries have similar guidelines although in India we are yet to evaluate and/or formulate our own safety guidelines. Data obtained from volunteers has demonstrated that exposure for more than 20 min to SAR of 4 W/kg results in temperature rise in body by approximately 1°C. Although human body can tolerate such an increase, it is uncertain whether long lasting elevation of body temperature increases the risk of adverse effect. In order to prevent such increased risk, a safety factor of 10 has been affirmed yielding a basic restriction for SAR of 0.4 W/kg for workers (controlled environment). The average time for the determination of the whole-body SAR in all cases has been kept at 6 min, a value which is considered a reasonable estimate of the thermal equilibrium time (NRPB, 1999). For general public (which may comprise subgroups of greater sensitivity like infants, the aged, the ill, and the disabled, potentially greater exposure durations as well as psychological/emotional factors ranging from anxiety to ignorance) an extra safety factor of 5 has been introduced, resulting in a maximum SAR of 0.8 W/kg (uncontrolled environment).

For therapeutic purposes it has been recommended that RF exposure should not be higher than that which could cause an increase in body temperature of more than 0.5°C and of any part of the body of more than 1°C. These limits can be satisfied if the SAR does not exceed 1 W/kg as averaged over 25% of the whole-body mass for exposures of longer than 15 minutes duration and 2 W/kg as averaged over any 25% of the whole-body mass for exposures of up to 15 min duration. Revised US FDA guidelines (1997) recommend the following SAR limits: 4 W/kg averaged over the whole body for any period of 15 minutes, 3 W/kg averaged over the head for any period of 10 minutes and 8 W/kg in any gram of tissue in the head or torso or 12 W/kg in any gram of tissue in the extremities, for any period of 5 minutes. This seems reasonable and is especially important for target tissues having limited capacity for heat dissipation due to limited blood flow such as the eye.

International Radio Wave Exposure Standards

Various countries have developed their own guidelines of safety limits including those developed by the U.S. National Council on Radiation Protection and Measurement (NCRP, 1986), the International Radiation Protection Association (INIRC-IRPA, 1988), the World Health Organization (WHO, 1993), International Electrical and Electronic Engineers (IEEE, 1999), the Canadian Safety Code 6 (1999) (Table 1). Of the various observed responses to RF radiation exposure which include changes in temperature regulation, endocrine function, cardiovascular function, immune response, nervous system activity and behaviour (Roberts et al 1986; Elder 1987; Cleary 1990), the behavioral responses have been considered to be among the most sensitive in the whole organism, and thus, of the greatest importance in setting guidelines for human exposure (WHO 1993; IEEE 1999). While developing these guidelines, a threshold exposure level of 4 W/kg for potentially adverse effects, has achieved a broad consensus, which is based predominantly on short-term behavioral studies in several species (D'Andrea, 1991). This is, however, not an unequivocal demarcation since some responses to thermalizing RF exposure at levels in the 1-2 W/kg range have been noted that are similar to those observed at or above 4 W/kg (Adair and Adams, 1980; DeLorge, 1984; Lotz, 1985).

Table 1.   International Radio Wave Exposure Standard (Source: Firstenberg A)

Country	Exposure Levels (µW/cm²)
New South Wales, Australia	0.001
Russia	2–10
Bulgaria	2–10
Hungary	2–10
Switzerland	2–10
China	7–10
Italy	10
Auckland, New Zealand	50
Australia	200
New Zealand	200–1000
Japan	200–1000
Germany	200–1000
United States	200–1000
Canada	200–1000
United Kingdom	1000–10,000

Current standards based on thermal effects in 10-300 GHz frequency range were reviewed in a WHO workshop conducted in the spring of 2002 on "Adverse temperature levels in the human body"; a consensus was reached that standards should consider both temperature and time of exposure wherever possible.

Biological Effects of Radiofrequency Radiation Exposures

Thermal Effects

All established hazards of RF radiation (RFR), the RFR injuries occur at exposure levels that cause heating of the body tissues by the absorption of the incident energy. A substantial database is available pertaining to evaluation of biological effects of RF exposures which are thermal in nature (i.e., exposures which deposit enough RF energy into the body to alter body temperature, and/or stimulate thermoregulatory responses). Changes in thermoregulatory activity have been shown to indirectly affect biological responses to RFR. Experiments pertaining to thermal exposures have generally been carried out on animals which were subjected to controlled RF exposures for short times of less than 8 hours and these publications, in fact, form the basis upon which various exposure guidelines have been formulated.

A sustained whole body exposure of rats to 35 GHz has been shown to induce oxidative stress which is preceded by circulatory failure (Kalns et al 2000). During 94 GHz radiation, cardiorespiratory changes by microwave induced lethal heat stress and beta adrenergic blockade have been reported in rats (Jauchem and Frei 1994; Jauchem et al. 1999). Human beings have an excellent capacity for maintaining thermal homeostasis, although it may not be improbable that high intensity exposures for the RFR induced heating may exceed the body's ability to dissipate heat which may result in temperature elevation capable of destroying a tissue; particularly vulnerable would be the tissues having poor blood perfusion namely the lens of the eyes. Human threshold for detection of 94 GHz MMW (based on a 10s exposure) is 4.5 mW/cm^2 (Blick et al 1997). Later modeling suggested that this detection threshold was based on a less than 0.1°C temperature rise, assuming a baseline temperature of 34°C (Riu et al 1997). The human threshold for pain perception to 94 GHz MMW is 1250 mW/cm^2, corresponding to a temperature rise of about 9.9°C (Walters et al 2000). Although damage thresholds have not been reported for MMW, it has been suggested that thermal damage occurs after a temperature rise of about 34°C (Ryan et al 2000, Walters et al 2000). Corneal sensitivity to thermal pain occurs at about 38-39°C in humans (Beuerman and Tanelian 1979). It is probable that high intensity MMWs act on human skin and cornea in an orderly, dose-dependent manner; detection occurs at very low power densities, followed by pain at higher exposures, followed by physical damage at even higher levels; exposure duration is a critical factor as the effective stimulus is joule heating. Recent reports indicating that the U.S. military is considering the use of 95 GHz MMW in a unique, non-lethal force protection application wherein the RF energy would be beamed onto humans at a distance in a controlled manner, so as to raise the skin temperature to a level that is painful but not damaging (Dao 2001) should be treaded with caution. Temperature elevation, if sufficiently high, is well known to cause a number of pathologic consequences (Miller and Ziskin 1989); proteins begin to denature and coagulate at sustained temperatures above 43°C which may lead to cell membrane rupture and cell death. Temperatures above 60°C can produce clinically recognizable burns and the amount of damage will depend on the magnitude and duration of the temperature elevation. The major difference between ordinary burns and the thermal damage from RFR is in the distribution of the damage. In ordinary burns from non-RFR sources and commonly encountered RF burns, heating occurs at surface of skin. Some frequencies of RF energy can penetrate deeper into the body and heating can occur at places where conventional surface heating cannot reach and this kind of internal heating can be greater than at the skin surface (Ziskin 2002). The medical consequences of thermal injury depend on the tissue type, their perfusion and capacity for regeneration; most vulnerable are the eyes and the nervous system (Lin 1993) neither of which regenerate. Heating of the testis can cause a temporary spermatopenia. The acute symptoms of nausea, vomiting, headache, loss of equilibrium, malise and fatigue are related to heating of central nervous system (Ziskin 2002). Visual blurring, cataract formation, dryness in mouth and palpitations may be long term effects of high exposures. Regarding the cancer promotion and co-promotion potential of 94 GHz MMW under acute (10 s exposure, 1000 mW/cm^2) or repeated (10 s exposure twice weekly for 12 weeks, 333 mW/cm^2) conditions, no effect of the MMW exposure were seen on SENCAR mouse model of skin cancer (Mason et al 2001).

The biological effects of low-intensity millimeter waves (MMW), 30-300 GHz, have been studied for decades in the countries of the former Soviet Union where MMW utilized extensively as a therapeutic modality (Akyel et al 1998; Rojavin and Ziskin, 1998; Pakhomov and Murphy 2000). MMW have been reported to produce a variety of biological effects, many of which are unexpected from a radiation penetrating less than 1 mm into biological tissue and could be termed as "nonthermal" MMW effects. Reported MMW induced biological effects include growth rate modulation and cell membrane effects (Pakhomov et al 1998), genotropic effects such as increased λ-phase (Lukashevsky and Belyaev 1989), induction and colicin synthesis (Pakhomov et al 1998) and enhanced DNA repair (Rojavin and Ziskin 1995). Belyaev et al (2000) observed change in the radial migration of DNA-protein complexes in a hydrodynamic field which were attributed to conformational change in the fine structure of both the eukaryotic and prokaryotic cell chromatin and changes in the binding of chromatin-associated proteins (Belyaev et al 1993). These changes are also shown to be dependent on a range of physical, physiological and genetic factors as well as have a strong frequency dependence (referred to as a "window" or "resonance" effect). Multiple windows have been reported in the 37–78 GHz mm-wave spectral range (Rojavin and Siskin 1998). MMW have been reported to produce a variety of bioeffects, many of which would be unexpected from a radiation which can penetrate only a less than 1 mm into biological tissues and could be termed as 'non thermal' MMW effects.

Thermal Exposures to Humans from Diagnostic and Therapeutic Devices

In interstitial thermal therapies which use electromagnetic energy (e.g. 344 MHz, 915 MHz) via an interstitial antenna (Couglin et al 1983) or external applicator arrays (Diederich et al 1991), or lower radiofrequency radiation e.g. 27 MHz (Delannoy et al 1990; Hall et al 1990) for treatment or ablation of benign and malignant lesions, the local SAR is usually 1000 W/kg which exceeds limits set by any international agency. RF therapy using ELF modulation of a 27 MHz carrier wave, which has recently been awarded FDA approval for the treatment of chronic psychophysiological insomnia, uses a device which has a maximum output power of 100 mW. This therapy is effective although the maximum SAR claimed is below safety limits variously defined for the general public (Pasche et al 1996) thus providing strong evidence that RF fields can elicit biological effects even below the prescribed safety limits. The greatest source of exposure to patients from RF comes from magnetic resonance imaging (MRI) devices (primarily around 60 MHz, but varying from 4–80 MHz). MRI operators are generally protected from RF exposure because the RF field drops off as one moves away from the RF transmit coil. Moreover, the patient and operator are usually separated by an efficient RF screen.

Nonthermal Effect

Effect on Cell proliferation and Cell Kinetics: The effects of radiofrequency fields on cell proliferation are nonthermal (Velizarov et al 1999). Cleary and his colleagues carried a series of experiments on cell proliferation and cell kinetic studies under continuous wave RF exposures at both 2.45 GHz and 27 MHz frequencies (SAR 5-50 W/kg) and showed increased proliferation under such fields (Cleary et al 1990a). No threshold for the effect was found and statistically significant differences were observed even at the lowest SAR (5 W/kg). Similar effects were also seen in human peripheral lymphocytes (Cleary et al 1990b). Alterations in cell-cycle kinetics under similar exposure conditions were also reported in Chinese hamster ovary cells (Cleary et al 1996). Stagg et al (1997) reported increased DNA synthesis in a glioma cell line (C6) as well as primary rat glial cells exposed to RF signals identical to certain cellular telephone signals (836.55 MHz, time domain multiple access-TDMA, SAR 5.9 mW/kg) for a longer period (24 h) but at lower exposure level than those used by Cleary et al (1996). In another study using RF exposures similar to cellular telephone signals from Global System for Mobile Communications (GSM), Kwee and Raskmark (1998) evaluated cell proliferation in cultures of transformed human epithelial amnion cells (AMA) exposed to 960 MHz at SARs of 0.021, 0.21 and 2.1 m W/kg for exposure times of 20, 30 or 40 min. A decrease in cell growth was seen at all three SAR levels tested, but only for exposures lasting 30 min or longer.

Microwave irradiation has been shown to down-regulate gap junctional intercellular communication (GJIC) which plays an essential role in regulation of cell growth, differentiation and wound healing) and the effect is strongly influenced by modulation frequency. It has been indicated that the mechanism of GJIC inhibition by ELF MF may be mainly due to hyperphosphorylation of gap junctional connexins by PKC rather than its transcriptional or translational disregulation (Chiang 1998).

Effect on Ca^{2+} Efflux: Increased $^{45}Ca^{2+}$ efflux from neonatal chick neural tissue, *in vitro* have been reported after exposure to an extremely low frequency modulated 147 MHz radiofrequency carrier with incident power density 10–20 W/m^2 (Bawin et al 1975). Such effects were absent in exposures under carrier frequency alone and peaked around 11–16 Hz modulation. However, in follow-up experiments using a 450 MHz RF carrier modulated at 16 Hz, a 1 W/m^2 and 10 W/m^2 power density increased $^{45}Ca^{2+}$ efflux, whereas 0.05 W/m^2 and 20 W/m^2 exposures were not effective in showing any increase (Bawin et al 1978). Such frequency modulated window phenomena has also been reported at 50 MHz (Blackman et al 1980a), 147 MHz (Blackman et al 1980b; Dutta et al 1989), 450 MHz (Sheppard et al 1979), and 915 MHz (Dutta et al 1984). Experiments conducted by Shelton and Merritt (1981) (1000 MHz carrier modulated at 16 Hz with 5, 10, 20, 150 W/m^2 and 32 Hz at 10, 20 W/m^2) in rat brain tissue, however, did not show any effect. Merritt et al (1982) also did not observe any effect in microwave (1000 MHz, 0.29 or 2.9 W/kg; 2450 MHz, 3 W/kg; 2060 MHz, 0.12 or 2.4 W/kg) irradiated rat brain tissue loaded with $^{45}Ca^{2+}$ by intraventricular injection. Exposure to 50 MHz radiofrequency (RF) non-ionizing radiation modulated (80%) with 16 Hz frequency has been reported to lead to overexpression of the ets1 mRNA in Jurkat T-lymphoblastoid and

Leydig TM3 cell lines. This effect was observed only in the presence of the 16 Hz modulation, corresponding to the resonance frequency for calcium ion with a DC magnetic field of 45.7 microT (Romano-Spica et al 2000).

Effect on Cell Membrane: RF fields have been reported to affect a variety of ion channel properties, such as decreased rates of channel protein formation, decreased frequency of single channel openings and increased rates of rapid, burst-like firing (these studies involved both CW and pulsed RF fields at a number of intensities; Repacholi 1998). Various studies have identified influences of MW exposure on Ca^{2+} release from cell membranes (Dutta et al 1984; Bawin et al 1975). Effects of RF/MW fields on transport of cations such as Na^+ and K^+ across cell membranes have also been documented and it has been suggested that these effects may occur without measured changes in temperature (Cleary 1995). An increase in calcium dependent protein kinase C has been noted in developing rat brain indicating that this type of radiation could affect membrane bound enzymes associated with cell signaling, proliferation and differentiation (Paulraj and Behari 2004). These effects have been reported to occur over a wide range of SARs (0.2-200 W/kg) and frequencies (27 MHz to 10 GHz). Any possible mechanism for the effect of RF/MW fields on the membrane channels or on the molecular structure of proteins or membrane lipids remains unresolved. Free radicals have been proposed to participate in RF-induced phase transitions in lipid vesicles exposed to CW fields at 0.2 W/kg (Phelan et al 1992). Microwaves have been shown to affect the kinetics of conformational changes of the protein betalactoglobulin and it can accelerate conformational changes in the direction towards the equilibrium state which applies both for the folding and the unfolding processes (Bohr and Bohr 2000).

Effect on Blood Brain Barrier: An increase in the blood brain barrier permeability in response to exposure to RF field have been reported in a number of studies (Albert 1977; Oscar and Hawkins 1977; Fritze et al 1997). Increased blood-brain barrier permeability in rats to a rhodamine-ferritin complex at 2450 MHz (100 W/m^2, 2 W/kg) have been observed (Neubauer et al 1990) although at a lower power density (5 m W/m^2) and a shorter time (15 minutes compared to 30 minutes) such effect was not seen. Changes in blood-brain barrier permeability were not seen if temperature increases in the brain were reduced by water cooling of the transmit antenna (Moriyama et al 1991). Fritze et al (1997) reported no significant increase in extravasation of serum proteins after rats were exposed to 9000 MHz microwaves modulated at 21 MHz at SAR values of 0.3 and 1.5 W/kg although at a SAR of 7.5 W/kg, a significant increase in serum albumin extravasations was seen.

With the advent of magnetic resonance imaging (MRI), the possibility of increased blood brain barrier permeability from RF exposure has been recently re-examined. Salford et al (1992, 1994) showed that microwave irradiation of rats (915 MHz, 0.016-5 W/kg, continuous wave or pulse modulated at 8, 16, 50, 200 Hz) significantly increased blood-brain barrier permeability to albumin in all exposed groups. No significant differences were observed between pulsed and continuous wave exposures. There was a significant increase in permeability for groups exposed to SAR values between 0.016 and 0.1 W/kg (whole body exposure), thus providing evidence that changes in blood-brain barrier permeability occur at SAR values below recommended safety limits. Oscar and Hawkins (1977), using much lower power densities than recommended as safe limits also showed increases in blood-brain barrier permeability. In a cell culture model, significant differences were observed in the BBB permeability in the exposed cultures compared to the controls (Schirmacher et al 1998). However, not all studies have shown significant increases, implying thereby that the changes may be related to the RF frequency or the extremely low frequency modulation of the RF carrier frequency. The possibility also exists that small increase in blood-brain barrier permeability may be initiated by the non-thermal effect of RF on ornithine decarboxylase activity or calcium ion concentrations (Koenig et al 1989).

Electrophysiological Effect: Exposure to high intensity RF fields which may be sufficient to cause a rise in temperature has been found to reduce the excitability of neurons (Arber and Lin 1984, 1985). Exposure to RF radiation at a high level has also shown a decrease in electrical amplitude of impulses and a reduction in excitability in frog sciatic nerve (McRee and Watchel 1982). However, at modest intensities such electrophysiological changes have not been noted (Chou and Guy 1978). Exposure to very low levels of amplitude modulated RF fields were reported to alter electrical activity in the brain of cats (Bawin et al 1973, 1974) and rabbits (Shandala et al 1979; Thuroczy et al 1994).

Effect on Neurotransmitters: Exposure to low-level pulsed and continuous wave (CW) RF has been reported to affect brain neurochemistry in a manner consistent with responses to stress (Lai et al 1988, 1989a). Interestingly, activation of cellular stress responses (hsp gene induction) was seen in transgenic nematodes (*Caenorhabditis elegans* strain PC72), carrying a stress-inducible reporter gene (*Escherichia coli* beta-galactosidase) under the control of a *C. elegans* hsp 16 heat-shock promoter (Daniells et al 1998) corroborating the fact that microwaves act as a "stressor". Acute high intensity microwave exposure of immobilized rats have been reported to induce minor stress responses which did not result in lasting adaptive or reactive changes of the brain (Fritze et al 1997). Modak et al (1981) reported a decrease in the concentration of acetylcholine in mouse brain under intense 2.45 GHz single pulse exposure. Dutta et al (1992) detected an increase in the activity of acetylcholinesterase in cultured human neuroblastoma cells exposed to low intensity RF field amplitude modulated at 16 Hz. These effects occurred over an SAR "window" but not at lower or higher intensities. Various changes in levels of norepinephrine and 5-hydroxytryptamine after acute and long-term exposure to RF have been reported (Snyder 1971; Inaba et al 1992). Changes in the amount of neurotransmitter substance released by nerve terminals could alter brain function.

Behavioural Responses: Disruption of operant behaviour under acute RF are well documented both in mice and primates (UNEP/WHO/IRPA, 1993).

Exposure of rhesus monkeys to RF field (resonant frequency 225 MHz wherein maximum energy is absorbed in them) resulted in reduced task performance at a whole-body SAR of 2.5 W/kg accompanied by a raised body temperature of about 1°C. Lai et al (1992, 1994) showed that a 45 min exposure to pulsed microwave fields affected radial-arm maze performance in rats and reduced high affinity choline uptake in frontal cortex and hippocampus. Wang and Lai (2000) also reported RF-induced changes in spatial memory. Impaired performance at 1°C rise in body temperature in rats at 600 MHz SAR 0.10–10 W/kg for 20 min have also been reported (Mickley et al 1994). Microwaves emitted by cellular telephones have been shown to modulate response patterns of human brain (Eulitz et al 1998) and also affects human slow brain potentials (Freude et al 2000). A decrease in expression of nitric oxide synthase (NOS) was noted in hippocampus of rats after exposure to electromagnetic pulse which probably related to the obstruction of learning and memory of the rat (Ding et al 1998). EMP exposure have been shown to result in changes of the content of neutrotransmitters in different cerebral areas of rats, lowering their ability of learning (Wu et al 1999).

Melatonin Studies: Stevens (1987) proposed a relationship between exposure to extremely low frequency (ELF) fields (50/60 Hz) and carcinogenesis through RF field action on melatonin secretion. Only a few studies have tested effect of RF exposure on melatonin synthesis. Stark et al (1997) studied dairy cattle herds located in the vicinity of short-wave (3-30 MHz) radio antenna and did not observe any chronic effect on salivary melatonin levels; however, a short term rise in melatonin was noted when the antenna was energised after being turned off for three days. In another study, no effect on noctural melatonin production was found (Vollrath et al 1997).

Genotoxic Effects: DNA rearrangement in a 7.7 kB *Hinf* I locus in cells from brain and testis were reported under RF field (2.45 GHz) at low intensity (SAR 1.18 W/kg) in Swiss albino mice (Sarkar et al 1994, 1996); the said 7.7 kB *Hinf* I locus could represent a hypermutable locus. Although it is not known whether exposure to a mutagenic agent or a specific class of mutagen increases the mutation rate in the region of these tandem repeats but it is known that stress induces amplification by extra replication of DNA segments in the non coding repeat sequences (Ramel 1989). RF may have downstream effect on cellular DNA and somatic intrachromosomal recombination inversion events have been noted (Sykes et al 2001). Increased dominant lethal mutations in the offspring of exposed male mice and abnormal sperm were also reported in Swiss albino mice (Verma et al 1976; Verma and Traboulay 1976; Goud et al 1982). Interestingly such effects were not seen in Sprague-Dawley rats (Berman et al 1980) or C3H mice (Saunders et al 1983, 1988). An increase in the number of single-strand and double-strand DNA breaks was reported in the brain cells of rats exposed for two hours to pulsed or continuous-wave 2.45 GHz radiation (Lai and Singh 1995, 1996) which were partially blocked by treatment with naltrexone indicating the endogenous opioids play a mediating role in RFR-induced DNA strand breaks in brain cells of the rat (Lai et al 1997). A few other studies have however not seen such damaging effects (Malayapa et al 1998).

Increased chromosomal aberrations have been reported in a large number of studies (Yao and Jiles 1970; Chen et al 1974, Garaj Vrhovac et al 1991, 1992; Khalil et al 1993; Maes et al 1993, 1995; Tice et al 2002). Chromosomal aberrations are generally thought to be due to damage to DNA and/or interaction between DNA and proteins and are usually associated with cancer and/or developmental abnormalities. Some studies have not noted such aberrations (Meltz et al 1987, 1989, 1990; Kerbacher et al 1990). Increased sister chromatid exchanges (SCEs is a phenomenon of switching of DNA from one part of the chromosome to another and is generally an indirect indicator of presence of genotoxic substance) have been reported (Khalil et al 1993; Maes et al 1997). Occurrence of increased micronuclei, which is another indirect indicator of DNA damage, have been reported in a large number of studies (Antipenko and Koveshinkova 1987; Maes et al 1993; Haider et al 1994; Balode 1996; Garaj-Vhrovac 1999).

Cumulative Effect of Radiofrequency Radiation

An important question regarding the biological effects of RFR is whether the effects are cumulative. This is particularly important while considering the possible health effects of mobile telephone usage, wherein repeated exposure of short duration over a long period (years) of time is involved. Existing results indicate changes in the response characteristics of the nervous system with repeated exposure and different types of outcomes have been reported e.g. an effect was observed only after prolonged (or repeated) exposure, but not after one period of exposure (Baranski, 1972; Baranski and Edelwejn 1975; Takashima et al 1979); the effect disappeared after prolonged exposure suggesting habituation (Johnson et al 1983; Lai et al 1992); and different effects were observed after different durations of exposure (Synder 1971; Baranski, 1972; Dumanski and Shandala, 1974; Grin 1974; Lai et al 1989b). Several lines of evidence suggest that responses of the central nervous system to RFR are cumulative and could represent a stress response (Lai et al 1987; Lai 1992). Stress effects are known to cumulate over time and involve first adaptation and then an eventual break down of homeostatic processes when the stress persists (Lai 1998). DNA damage in cells is also cumulative although Lai and Singh (1995) found that a single episode of RFR exposure increased DNA damage in brain cells of the rat.

Evidence of Possible Health Effects of Radiofrequency Radiation

Adult Cancer: There are few published studies on risk of adult cancers, childhood cancers, reproductive outcomes and congenital anomalies from RF exposure. Many of the studies of risk of cancer in adults with exposure to radiofrequency fields have been conducted in occupational groups, specially military personnel. A major review of epidemiological studies of RF exposure and cancer was recently published (Elwood 1999; www.wire.org/bibliography/expose.html). In January 1995, a report on epidemiological evidence of radiofrequency radiation was published focussing primarily on military, industrial and broadcast exposures (Goldsmith, 1995). This report was followed and supplemented by review of Rothman et al (1996). Goldsmith (1997) published another review on epidemiological evidence relevant to radar effects. Robinette et al (1980) reported on a study of two cohorts of US enlisted naval personnel exposed during Korean War and followed for about twenty years. The study noted frequent lympho-hematological cancer and more disabilities of several causes in heavily exposed personnel compared to the less exposed men. Studies on hospitalisation of men with leukaemia by exposures in US navy were carried out by Garland et al (1990) and despite an inadequate period of follow up, those with most magnetic field exposures were found to have more leukemia.

Grayson (1996) reported on brain cancer in US Air Force personnel and found that nonionizing radiation particularly microwave exposure was more significant for positive association. Szmigielski (1996) reported several studies of cancer experience among Polish Army personnel depending on whether or not they had RF exposure. Overall the exposed personnel showed significantly more cancer like leukemia, lymphoma and brain tumor. A cluster of six cases of testicular cancer were reported among traffic policemen using microwave generators (Davies and Mostofi 1993). Hayes et al (1990) reported excess risk of testicular cancer among military personnel who self-reported exposure to microwaves and radio waves.

Muhm (1992) reviewed mortality of male employees of an electromagnetic pulse (EMP) test program using death certificate information. Armstrong et al (1994) examined exposure to pulsed electromagnetic fields in a nested case-control study of utility workers and showed excess risk of lung cancer. Lin et al (1985) collected data on brain cancer deaths among white males in Maryland occupationally exposed to electromagnetic fields and showed greater association of cancer among the electric, electronic engineers and technicians. Milham (1985) analyzed mortality experience and reported increased mortality from leukemia and non-Hodgkin's lymphomas in workers presumably exposed to electric and magnetic fields including radio and telegraph operators and repairmen. Thomas et al (1987) reported an increased risk of brain tumor death in men ever employed in an electronics occupation although this was independent of microwave/radiofrequency exposure. Tynes et al (1996) reported increased breast cancer risk among female radio and telegraph operators. In a study of mortality experience of a cohort of Italian plastic-ware workers exposed to radiofrequency fields generated by dielectric heat sealers, it was observed that mortality from malignant neoplasms was slightly elevated, and increased risks of leukemia and accidents were detected (Lagorio et al 1997).

Community studies on cancer associated with possible proximity to broadcast facilities have been reported (Hocking et al 1996; Dolk et al 1997a, b). Leukemia mortality was found to be higher than expected near a high power radio-transmitter in a peripheral area of Rome (Michelozzi et al 1998).

Childhood Cancer: Children of US embassy personnel residing in Moscow and East European Embassies, who were irradiated by Soviet Union in the 1960s, showed increased incidence of leukemia while chromosomal aberrations were more frequent in the exposed personnel (Lilienfeld et al 1978). "Prof. Lilienfeld's report was never actually published and at last when it did become available to public, all the statements of concern made by him were removed and a modified report was taken out showing no serious health effect" (Goldsmith, 1997).

Reproductive Outcomes: Studies on gonadic function in young men with exposure to microwaves (in the range 10 s to 100 μW/cm^2 showed alteration of spermatogenesis (Lacranjan et al 1974). Semen analysis of military personnel associated with potential microwave exposure showed lower sperm count than comparison group (Danulescu et al 1996; Weyandt et al 1996; Schrader et al 1998). Differences in semen quality and hormone levels have also been observed in RF dielectric heater operators (Grajewski et al 2000).

In a retrospective study of physiotherapists, it was found that those who worked during early pregnancy with microwave equipment had a significantly greater incidence of spontaneous abortion than if they worked with short wave diathermy equipment (Oullet-Hellstrom and Stewart 1993).

Haematological Effect: Nearly all blood counts among personnel exposed to irradiation in Moscow were different than those taken in Washington DC (Lilienfeld et al 1978). Steneck et al (1980) noted that the haematological changes were among the first health effect to be found in personnel exposed to RF. Daily (1980) reported a statistically significant increase in immature red blood cells among workers exposed to radar when the same was first identified as a health risk. Goldoni (1990) compared the hematological findings in 25 male air traffic control technicians working at a distance from microwave source and reported that radar exposed workers had significantly lower levels of leukocytes and red cells than the electronic technicians. In a follow up study of 49 radar exposed technicians, thrombocyte and leucocyte counts decreased though they stayed within normal limits (Goldsmith 1997). Electrocardiographic abnormalities were detected significantly more frequently ($p = 0.006$) in workers exposed to electromagnetic fields than in non-exposed subjects (75% versus 25%) and a higher number of rhythm disturbances was observed in AM broadcast station workers (Bortkiewicz et al 1997).

DNA Sequence Rearrangement/Mutation: Evidences of mutational activity in human incubated white blood cells have been reported from personnel who were accidentally exposed while repairing microwave devices (Garaj-Vrhovac et al 1993; Garaj-Vrhovac 1999) though such effects have not been reported in normal lymphocytes under low frequency low energy pulsed electromagnetic field (Emilia et al 1985). Garaj-Vrhovac et al (1992) examined RF radiation effect on human lymphocytes and showed correlation between micronuclei percentage and specific chromosomal aberrations. Genotoxic effects of amplitude modulated microwaves was also found in human lymphocytes in culture (d'Ambrosio et al 1995, 2002).

Effect of microwave radiation on human DNA was studied in personnel working with instruments known to emit microwave radiation using synthetic oligo probes in conjunction with different restriction enzymes (Sarkar and Ali 1999). Most important observations were made in the *Bam*HI digested DNA in the exposed worker wherein the *Bam*HI digest uncovered sequence modulation in one of the alleles in the region of 12-13 kb in the personnel with increased frequency (53.17%) compared to the controls (26.31%) suggesting that some loci in the human DNA are more prone to mutation owing to electromagnetic field. Although this observation does not provide a conclusive evidence that the appearance of the 12-13 kb band is due to radiofrequency radiation, but a two-fold increased frequency in exposed individuals seems to be an attractive possibility (Sarkar 2001a). Analysis of a minisatellite region representing the consensus sequence of the human hypervariable locus 033.15 (Jeffreys et al 1985) under radiofrequency field, however, showed no apparent difference in the band profile in the exposed workers vis-á-vis the controls implying thereby that probably the alleles in this locus are not affected by radiofrequency radiation (Sarkar 2001b). Status of the DYZ1 fraction was also ascertained in the germline (semen) DNA of some of the exposed workers. Though the profile in both the somatic and germline sample was the same, it was interesting to note the presence of a band in the region of 500 bp in the semen sample which was absent in the DNA from blood (Sarkar and Selvamurthy 2001; Sarkar 2001b). Significance of the alteration of the restriction site for *Rsa*I enzyme in the DYZ1 locus in the germinal tissue is not clear yet.

Psychological Effect: Unfavourable influence of chronic occupational exposure to pulsed microwave have been observed on psychological reactivity reducing the individual resistance to stressful events, increasing the level of tiredness, reaction time, troubles of recent memory as also inducing an excitatory state that can subsequently induce inhibition and tiredness (Balaceanu and Danulescu, 1996).

Evidence of Other Health Effects: Other health outcomes reported after RF exposure include headache, general malaise, short-term memory loss, nausea, changes in EEG and other central nervous system functions and sleep disturbances (von Klitzing 1995; Reiser et al 1995; Altpeter et al 1995; Bortkeiwicz et al 1995; Mann and Roschke 1996). Hypersensitivity to electromagnetic field with symptoms of headache, insomnia, tingling, skin rashes, dizziness and lack of concentration are some of the other effects reported after radiofrequency exposure (Berg et al 1992; Bergqvist and Wahlberg 1994; Berdahl 1995; Sandstrom et al 1995). Disturbances in glucose tolerance and disorder in bioelectrical activity of brain have been reported in workers exposed to electromagnetic radiation (Bielski and Sikorski 1996). Alteration of diurnal rhythms of blood pressure and heart rate in workers exposed to radiofrequency electromagnetic fields have been reported (Szmigielski et al 1998).

Published laboratory studies with animal experimentation that relate to cancer causation and other bio-effects are also available (RSC.EPR 99-1). Radiation specific cataract has also been found in personnel working with extremely high microwave exposure (Toncheva et al, quoted in Goldsmith, 1997). Exposure of the right hemisphere to a radiofrequency EMF for 35 min caused an increase in sympathetic effect activity in human subjects which increased the resting blood pressure between 5 and 10 mmHg. The effect was likely to be caused by the vasoconstriction (Braune et al 1998).

Radiowave Sickness: A large number of clinical studies have reported radiowave sickness in workers exposed on the job (Sadchikova 1960, 1973, 1980; Klimkova-Deutschova 1974; Baranski and Edelwejn 1975; Zalyubovskaya and Kiselev 1978; Bachurin 1979; Gorbach 1982; Trinos 1982; Markarov et al 1995; Liakouris 1998). The term "radio wave sickness" was first used by Russian doctors to describe an occupational illness developed by large numbers of workers exposed to microwave or radiofrequency radiation. The symptoms were called "neurasthenic" and included insomnia, headaches, dizziness, nausea, memory loss, difficulty in concentrating, irritability, respiratory illness (bronchitis, sinusitis, pneumonia), flue-like illness, asthma, fatigue, weakness, pressure or pain in the chest, increase in blood pressure, altered pulse rate (usually slowed), pressure behind the eyes, other eye problems, swollen throat, dry lips or mouth, dehydration, sweating, fever, shortness of breath, muscle spasms, tremors, pain in the legs or the soles of the feet, testicular or pelvic pain, joint pain, pains that move around the body, nosebleeds, internal bleeding, hair loss, digestive problems, skin rash, ringing in the ears, impaired sense of smell, pain in the teeth (especially with metallic fillings) (Firstenberg, 1997).

Effect of Pulsed Electromagnetic Radiation

Psychotropics: People having normal hearing perceive pulse modulated RF fields with carrier frequencies between ~ 200 MHz and 6.5 GHz. This is referred to as the "microwave hearing effect," a phenomena discovered in the 1960s by Frey. The sound has been variously described as a buzzing, clicking, hissing or popping sound depending on modulation characteristic. According to Frey (1963), if the millimeter waves have enough energy density and are powerful enough, there are other phenomena where one could cause sort of a concussion kind of effect which could conceivably be heard by bone conduction. It would transfer through skin to bone and bone into the inner ear. It is possible to modulate such energy to create the perception of some intelligible sounds. U.S. military proponents of the non-lethal ray describe goals of using electromagnetic radiation for such mind control.

It also appears possible to create high fidelity speech in the human body, raising the possibility of covert suggestion and psychological direction. When a high power microwave pulse in the gigahertz range strikes the human body, a very small temperature perturbation occurs. This is associated with a sudden expansion of the slightly heated tissue. This expansion is fast enough to produce an acoustic wave. If a pulse stream is used, it should be possible to create an internal acoustic field in the 5-15 kilohertz range, which is audible. Thus, it may be possible to 'talk' to selected adversaries in a fashion that would be most disturbing to them. Exposure to ultrawideband (UWB) electromagnetic pulses, which are being used as a new modality in radar technology, has been shown to cause hypotension in male Wister rats (Lu et al 1999). It has also been demonstrated that pulsed high-frequency EMF in the range of radiotelephones may promote sleep and modify the sleep EEG (Borbely et al 1999).

Electromagnetic Pulse (EMP) Weapons: The explosion from EMP weapons can generate high power microwaves that can disable electronic circuitry in computers and communication equipment. It has also been found that high altitude burst can produce large amplitude EMP field over thousands of kilometers and peak energy fields can reach levels of 50 kilovolts per meter very quickly and would have large broadband frequency coverage. The biological effect of RFR exposure emitted from such weapon systems is yet to be ascertained.

Blame it on the Cellular Phone!

In the 1990s the explosive development of cellular phone systems has greatly increased the extent and magnitude of RFR exposure. These new telecommunications technologies have been introduced without full provision of information about their nature and without prior discussion within the scientific community about possible consequences for health. Potential exposure also develop in the vicinity of the fixed broadcast facilities often located in residential areas, schools etc. with no attention to possible harmful exposure of occupants nearby. Exposure to RFR from mobile telephones is of a short-term, repeated nature at a relatively high intensity, whereas exposure to RFR emitted from cell masts is of long duration but a very low intensity. As the costs of mobile phone technology have fallen, their use has increased dramatically and the overall levels of exposure of the population as a whole have therefore increased. Mobile telephone radiation is modulated at low frequencies and it is established that modulated or pulsed RFR are more effective in producing an effect. They also elicit different effects when compared with continuous-wave radiation of the same frequency (Baranski 1972; Frey and Feld 1975; Frey et al 1975; Oscar and Hawkins 1977; Arber and Lin 1985; Sanders et al 1985; Lai et al 1988). This also raises the possibility of biological effects of low frequency electric and magnetic fields of mobile telephone radiation. Biological effects of low frequency (< 100 Hz) electric and magnetic fields are quite well established. Of somewhat greater concern is the exposure to the user of hand held phones, especially those with no external antenna. A number of such phones were shown to produce thermal absorption in the head of the user which even existing standards found excessive. The close proximity of a mobile telephone antenna to the user's head leads to the deposition of a relatively large amount of radiofrequency energy in the head. The relatively fixed position of the antenna to the head causes a repeated irradiation of a more or less fixed amount of body tissue (van Leeuwen et al 1999). 'Hot spots' may form in certain areas of the head. As a reference, from theoretical calculations (Dimbylow 1993; Dimbylow and Mann 1994; Martens et al 1995), peak (hot spot) SAR in head tissue of a user of mobile telephone can range from 2 to 8 W/kg per watt output of the device. The peak energy output of mobile telephones can range from 0.6-1 watt, although the average output could be much smaller. Chou et al (1985) measured local energy absorption rates (SARs) in different areas of the brain in a rat exposed to RFR and showed that two brain regions less than a millimeter apart could have more than a two-fold difference in SAR. The rat was stationary when it was exposed. The situation is more complicated if an animal is moving in an RF field. Depending on the amount of movement of the animal, the energy absorption pattern in its body could become either more complex and unpredictable or more uniform. Testicular function of rats have been found to be affected by whole-body microwave exposure emitted by cellular phones (Dasdag et al 1999). The average output power from the antennas of digital mobile phones is lower than that from earlier analogue models, but the maximum powers are greater, the exact patterns of radiation are different and these differences might influence their effects on people. A study conducted in Sweden on about 209 brain tumors suggested development of cancer in head region close to the use of cell phones (Hardell et al 1999). Study of mobile phone users have shown a statistically significant association between calling time/number of calls per day and the prevalence of warmth behind/around the ear, headaches, and fatigue (Hanson et al 1998). A number of studies have reported effects on brain function under cellphone usage (www.wire.org/bibliography/expose.shtm). It has been shown that occupational cellular telephone use may be associated with reduced daytime melatonin production (Burch et al 1997). Mobile phones have been shown to modulate response patterns of human brain activity (Eulitz et al 1998) and also affect human slow brain potentials (Freude et al 2000). A 15-min exposure to GSM phone radiation caused an increase in auditory brainstem response in the exposed side of human subjects. Subjects also showed a hearing deficiency in the high frequency range (20 dB hearing deficiency from 2 KHz to 10 KHz) (Kellenyi et al 1999). Exposure to EMF emitted by cellular phones do not alter the resting EEG *per se* but modifies the brain responses significantly during a memory task (Krause et al 2000).

Symptoms and signs which included pain, headache, numbness, and parasthesiae, malaise, diarrhoea, and skin erythema were reported from antenna engineers who were accidentally exposed to high levels of ultrahigh frequency radiofrequency radiation (785 MHz mean frequency) while working on a television mast. The most notable problem was that of acute then chronic headache involving the part of the head which was most exposed (Schilling 1997, 2000). Results of epidemiological studies in UK and Australia suggested that exposure of residents living near (i.e. within 5 kms) the broadcast facilities may have small increase in leukemia (Dolk et al 1997a, b; Hocking et al 1996).

Current and Future Research

Many research programs are currently underway throughout the world focusing on the relationship between mobile phones and health viz., the WHO International EMF Project (start date 1997, target completion date 2005). International collaborators in this project are: International Commission on Non-Ionizing Radiation Protection (ICNIRP), International Agency for Research on Cancer (IARC), International Labour Organization (ILO), European Commission (EC), International Telecommunication Union (ITU), International Electrotechnical Commission (IEC), United Nations Environment Programme (UNEP), North Atlantic Treaty Organization (NATO) and over 40 governmental agencies. Independent Collaborators are: National Radiological Protection Board (UK), Bundesamt Fur Strahlenschutz (Germany), Karolinska (Sweden), Food and Drug Administration (USA), National Institute of Environmental Health Sciences (USA), National Institute of Occupational Health (USA) and National Institute for Environmental Studies (Japan). In the last week of June 2000, WHO gave a new set of recommendations for the safe use of mobile phones asking consumers to limit the exposure to harmful radiofrequency by limiting length of calls or using hand-free devices thus keeping the instrument away from head and body. It also recommended strict adherence to international guidelines by not just mobile phone users but also those who work near or live around base station. Furthermore, setting base station near kindergarten, schools and playground may also need special consideration.

Conclusion

Health, as defined by WHO, is a state of complete physical, mental and social well-being and not merely the absence of disease or infirmity. A hazard implies an effect that the subject would regard as adverse to his or her well being. Evidence of health hazard will depend on whether the effects are reversible or are within range for which effective compensation mechanism exist or are likely to lead to unfavourable effects on physical, mental and social well being. Measurable biological and psychological changes have been found to be associated with RF exposure; of somewhat greater concern would be the non reversible genotoxic effects of radiofrequency radiation. There also exists a carcinogenic potential and other health effects associated with radiofrequency exposure. If cancer risk cannot be unequivocally demonstrated today, still it will be advisable and prudent to adhere to some measure of protection as neoplastic responses have a long latency period. Radiofrequency radiation may not be assumed to be harmless and innocent.

Acknowledgement

The work was carried out under DRDO grant no. INM-222 and DIP-213 to S.S. The authors are thankful to Ms Shuchi Sharma for typing assistance.

References

Adair ER, Adams BW. Microwaves induce peripheral vasodilation in squirrel monkey. *Science* **207**:1381–1383, 1980.

Akyel Y, Pakhomova ON, Stuck BE, Murphy MR. Current State and Implications of Research on Biological Effects of Millimeter Waves: A Review of the Literature. *Bioelectromagnetics* **19**:393–413, 1998.

Albert EN. Light and electron microscopic observations on the blood-brain barrier after microwave irradiation. In Symposium on Biological Effects and Measurements of Radiofrequency Microwaves Washington DC: DHEW (HEW Publications), FDA77-8026, pp 294–304, 1977.

Altpeter ES, Krebs TH, Pluger DH, et al. Study on health effects of shortwave transmitter station of Schwarzenberg. Berne, Switzerland University of Berne. BEW Publication Series. Study No. 55. Berne. pp. 155, 1995.

Antipenko EN, Koveshnikova IV. Cytogenetic effects of microwaves of non-thermal intensity in mammals. *Dokl Akad Nauk SSSR* **296**: 724-726, 1987.

Arber SL, Lin JC. Microwave enhancement of membrane conductance: Effects of EDTA, caffeine and tetracaine. *Physiol Chem Phys Med NMR* **16**: 469–475, 1984.

Arber SL, Lin JC. Microwave-induced changes in nerve cells. Effects of modulation and temperature. *Bioelectromagnetics* **6**: 257-270, 1985.

Armstrong B, Theriault G, Guenel P, Deadman J, Goldberg M, Heroux P. Association between exposure to pulsed electromagnetic fields and cancer in electric utility workers in Quebec, Canada and France. *Am J Epidemiol* **104**: 805–20, 1994.

Bachurin IV. Influence of small doses of electromagnetic waves on some human organs and systems. *Vrachebnoye Delo* **7**: 95–97, JPRS 75515, pp. 36-39, 1979.

Balaceanu G, Danulesco R. Specific features of psychological reactivity in professional exposure to pulsed microwaves. proc. International Congress on Radiation Protection (IRPA9), Vienna, Austria, Vol 3, pp 633, 1996.

Balode Z. Assessment of radio-frequency electromagnetic radiation by the micronucleus test in bovine peripheral erythrocytes. *Sci Total Environ* **180**: 81–85, 1996.

Baranski S. Histological and histochemical effect of microwave irradiation on the central nervous system of rabbits and guinea pigs. *Am J Phys Med* **51**: 182–91, 1972.

Baranaski S, Edelwejn z. Experimental morphologic and EEG studies of microwave effects on the nervous system. *Annals of the New York Academy of Sciences* **247**: 109–116, 1975.

Bawin SM, Gavalas-Medici RJ, Adey W.R. Effects of modulated very high frequency fields on specific brain rhythms in cats. *Brain Res* **58**: 365–384, 1973.

Bawin SM, Gavalas-Medici RJ, Adey WR. Reinforcement of transient brain rhythms by amplitude-modulated VHF fields. Biological and Clinical Effects of Low Frequency Magnetic and Electric Fields (J G Llaurado, A Sances and J H Battocletti, Eds). Springfield, Charles C Thomas p. 172, 1974.

Bawin SM, Kaczmarek LK, Adey WR. Effects to modulated VHF fields on the central nervous system. *ANN NY Acad Sci* **247**: 74–81, 1975.

Bawin SM, Sheppared A, Adey WR. Possible mechanisms of weak electromagnetic field coupling in brain tissue. *Bioelectrochem Bioenerg* **5**: 67-76, 1978.

Belyaev IY, Shcheglov VS, Alipov YeD, Radko SP. Regularities of separate and combined effects of circularly polarized millimeter waves on E. coli cells at different phases of culture growth. *Bioelectrochem Bioenerg* **31**: 49-63, 1993.

Belyaev IY, Shcheglov VS, Alipov ED, Ushalov VD. *IEEE Trans. Microwave Theory Tech* **48**: 2172, 2000.

Berman E, Carter HB, House D. Tests for mutagenesis and reproduction in male rates exposed to 2450-MHz (CW) microwaves. *Bioelectromagnetics* **1**: 65-76, 1980.

Blackman CF, Benane SG, Elder JA, House DE, Lampe JA, Faulk JM. Induction of calciumion efflux from brain tissue by radiofrequency radiation: Effect of sample number and modulation frequency on the power-density window. *Bioelectromagnetics* **1**: 35–43, 1980a.

Blackman CF, Benane SH, Joines WT, Hollis MA, House DE. Calcium-ion efflux from brain tissue: Power-density versus internal field-intensity dependencies at 50 MHz RF radiation. *Bioelectromagnetics* **1**: 277–283, 1980b.

Berg M, Arentz B, Lidin S, Eneroth P, Kallner A. Techno-stress: A psychophysiological study of employees with VDU-associated skin complaints. *J Occup Med* **34**: 698–701, 1992.

Bergdahl J. Psychologic aspects of patients with symptoms presumed to be caused electricity or visual display units. *Acta Odontol Scand* **53**: 304-310, 1995.

Bergqvist, Wahlberg J. Skin symptoms and disease during work with visual display terminal, 1994.

Beuerman RW, Tanelian DL. Corneal Pain Evoked by Thermal Sensation. *Pain* **7**: 1-14, 1979.

Bielski J, Sikorski M. Disturbances of glucose tolerance in workers exposed to electromagnetic radiation. *Med Pr* **47**: 227–231, 1996.

Blick DW, Adair ER, Hurt WD, Sherry CJ, Walters T J, Merritt JH. Thresholds of Microwave-Evoked Warmth Sensations in Human Skin. *Bioelectromagnetics* **18**: 403–409, 1997.

Bohr H, Bohr J. Microwave enhanced kinetics observed in ORD studies of a protein. *Bioelectromagnetics* **21**: 68–72, 2000.

Borbely AA, Huber R, Graf T, Fuchs B, Gallmann E, Achermann P. Pulsed high-frequency electromagnetic field affects human sleep and sleep electroencephalogram. *Neurosci Lett* **275**: 207–210, 1999.

Bortkiewicz A, Zmyslony M, Palczynski C, Gadzicka E, Szmigielski S. Dysregulation of autonomic control and cardiac function in workers at AM broadcasting station (0.738 - 1.503 MHz). *Electromagnetobiology* **14**: 177–192, 1995.

Bortkiewicz A, Zmyslony M, Gadzicka E, Palczynski C, Szmigielski S. Ambulatory ECG monitoring in workers exposed to electromagnetic fields. *J Med Eng Technol* **21**: 41-46, 1997.

Braune S, Wrocklage C, Raczek J, Gailus T, Lucking CH. Resting blood pressure increase during exposure to a radio-frequency electromagnetic field. *Lancet* **351**: 1857–1858, 1998.

Burch JB, Reif JS, Pitrat CA, Keele TJ, Yost MG. Cellular telephone use and excretion of a urinary melatonin metabolite. Abstract of the Annual Review of Research on Biological Effects of Electric and Magnetic Fields from the Generation, Delivery and Use of Electricity, San Diego, CA, pp. 110, 1997.

Canadian Safety Code 6. Limits of human exposure to radiofrequency electromagnetic fields in the frequency range from 3 kHz to 300 GHz. Ottawa: Environmental Health Directorate, Health Cannada, 1999.

Chen KM, Samuel A, Hoopingarner R. Chromosomal aberrations of living cells induced by microwave radiation. *Environ Lett* **6**: 37-46, 1974.

Chiang H. Microwave and ELF electromagnetic field effects on intercellular communication, *Proceedings of the 20th Annual International Conference of the IEEE Engineering in Medicine and Biology Society* **20**: 2798-2801, 1998.

Chou CK, Guy AW. Effects of electromagnetic field son isolated nerve and muscle preparation. *IEEE Trans Microw Theory Tech* **26(3)**: 141–147, 1978.

Chou CK, Guy AW, McDougall J, Lai H. Specific absorption rate in rats exposed to 2450-MHz microwaves under seven exposure conditions. *Bioelectromagnetics* **6**: 73–88, 1985.

Cleary SF. Biological effects of radiofrequency electromagnetic fields. In: Biological Effects and Medical Applications of Electromagnetic Energy. edited by Gandhi OP. Prentice-Hall, Englewood Cliffs, New Jersey, pp 236–255, 1990.

Cleary SF, Liu LM, Merchant RE. In vitro lymphocyte proliferation induced by radio-frequency electromagnetic radiation under isothermal conditions. *Bioelectromagnetics* **11**:47–56, 1990a.

Cleary SF, Liu LM, Merchant RE. Glioma proliferation modulated in vitro by isothermal radiofrequency radiation exposure. *Radiat Res.* **121**: 38–45, 1990b.

Cleary SF. Effects of radiofrequency radiation on mammalian cells and biomolecules in vitro. In Blank M. (ed.) "Electromagnetic Fields: Biological Interactions and Mechanisms". Washington: American Chemical Society, pp 467–477, 1995.

Cleary SF, Cao G, Liu LM. Effects of isothermal 2450 MHz microwave radiation on the mammalian cell cycle: comparison with effects of isothermal 27 MHz radiofrequency radiation exposure. *Bioelectrochem Bioenerget* **39**: 167–173, 1996.

Coughlin CT, Douple EB, Strohbehn JW, Eaton WL, Tremblay BS, Wong TZ. Interstitial hyperthermia in combination with brachytherapy. *Radiology* **148**: 285–288, 1983.

Daily LE. A clinical study of the results of exposure of laboratory personnel to radar and high frequency radio US naval medical Bulletin **41**: 1052–1056, 1943. Cited in Steneck NH, Cook HJ, Vander AJ, Kane GL. Origins of US safety standards for microwave radiation. *Science* **208**: 123–127, 1980.

Daniells C, Duce I, Thomas D, Sewell P, Tattersall J, Pomerai D de. Transgenic nematodes as biomonitors of microwave-induced stress. *Mutat Res* **399**: 55–64, 1998.

Danulescu E, Denulescu R, Popa D. Effect of radar occupational exposure on the male fertility, Proc International Congress on Radiation protection, Vienna, Vol 3, pp 632, 1996.

Dao, James, "Pentagon Unveils Plans for a New Crowd-Dispersal Weapon," New York Times, page A11, March 1, 2001.

D'Andrea JA: Microwave radiation absorption: behavioural effects. *Health Phys* **61**: 29–40, 1991.

Dasdag S, Ketani MA, Akdag Z, Ersay AR, Sari I, Demirtas OC, Celik MS. Whole-body microwave exposure emitted by cellular phones and testicular function of rats. *Urol Res* **27**: 219–223, 1999.

Davis RL, Mostofi FK. Cluster of testicular cancer in police officers exposed to hand-held radar. *Am J Industrial Med* **24**: 231–3, 1993 Aug.

Delannoy J, LeBihan D, Hould DI, Levin RL. Hyperthermia system combined with a magnetic resonance imagining unit. *Med Phys* **17**: 855–860, 1990.

De Lorge JO. Operant behaviour and colonic temperature of *Macaca mulatta* exposed to radiofrequency fields at and above resonant frequencies. *Bioelectromagnetics* **5:** 233-246, 1984.

Diederich CJ, Sherwin G, Adams C, Stauffer PR. Evaluation of a multi-element microwave applicator for hyperthermia. Proc 9th Int'l Conf Radiation Research, Toronto Canada, (unpublished): 194, 1991.

Dimbylow PJ. FDTD calculations of SAR for a dipole closely coupled to the head at 900 MHz and 1.9 GHz. Phys. *Med Biol* **38:** 361–368, 1993.

Dimbylow PJ, Mann, JM. SAR calculations in an anatomically realistic model of the head for mobile communication transceivers at 900 MHz and 1.8 GHz. Phys. *Med Biol* **39:** 1527–1553, 1994.

Ding G, Xie X, Zhang L, et al. Changes of nitric oxide synthase in hippocampus and cerebellum of the rat following exposure to electromagnetic pulse. *Chin J Phys Med* **20:** 81–83, 1998.

Dolk H, Shaddick G, Walls P, Grundy C. Thakrar B, Kleinschmitt I, Elliott P. Cancer incidence near radio and television transmitter in Great Britain. I. Sutton Coldfield transmitter. *Am J Epidemiol* **145:** 1-9, 1997a.

Dolk H, Ellott P. Shaddick G, Walls P, Thakrar B. Cancer incidence near radio and television transmitters in Great Britain. *Am J Epidemiol* **145:** 10–17, 1997b.

d'Ambrosio G, Lioi MB, Massa R, Zeni O, Scraft MR. Genotoxic effects of amplitude-modulated microwaves on human lymphocyte exposed in vitro under controlled condition. *Electro Magnetobiol* **14:** 157–164, 1995.

d'Ambrosio G, Massa R, Scrafi MR, Zeni O. Cytogenetic damage in human lymphocytes following GMSK phase modulated microwave exposure. *Bioelectromangetics* **23:** 7–13, 2002.

Dumansky, JD, Shandala Mg. The biologic action and hygienic significance of electromagnetic fields of super high and ultra high frequencies in densely populated areas, in: "Biologic Effects and Health Hazard of Microwave Radiation: Proceeding of an International Symposium, P. Czerski, et al., eds., Polish Medical Publishers, Warsaw, 1974.

Dutta SK, Subramanian A Ghosh B, Parshad R. Microwave radiation-induced calcium efflux from brain tissue in virto. *Bioelectromangetics* **5:** 71–78, 1984.

Dutta SK, Ghosh B, Blackman CF. Radiofrequency radiation induced calcium ion efflux enhancement from human and other neuroblastoma cells in culture. *Bioelectromagnetics* **10:** 197–202, 1989.

Dutta SK, Das K, Ghosh B, Blackman CF. Dose dependence of acetylcholinesterase activity in neuroblastoma cells exposed to modulated radio-frequency electromagnetic radiation. *Bioelectromagnetics* **13:** 317–322, 1992.

Elder JA. Radiofrequency radiation activities and issues: a 1986 perspective. *Health Phys* **53:** 607–611, 1987.

Elwood MJ. A critical review of epidemiologic studies of radiofrequency exposure and human cancers. *Environ Health Perpect* **107 (Suppl 1):** 155–68, 1999.

Emilia G. Torelli G, Ceccherelli G, Donelli A, Ferrari A, Zucchini P. PEMFs on the response to lectin stimulation of human normal and chronic lymphoctyic leukemia lymphocytes. *J Bioelect* **4:** 145-162, 1985.

Eulitz C, Ullsperger P, Freude G, Elbert t. Mobile phones modulate response patterns of human brain activity. *Neuroreport* **9:** 3229-3232, 1998.

Firstenberg A. *Microwaving Our Planet: The Environmental Impact of the Wireless Revolution*, Cellular Phone Taskforce, P.O. Box 100404, Brooklyn, New York 11210, or P.O. Box 1337, Mendocino, CA 95460, 1997.

Freude G, Ullsperger P, Eggert S, Ruppe I. Microwaves emitted by cellular telephones affect human slow brain potentials. *Eur J Appl Physiol* **81**: 18–27, 2000.

Frey AH. Human response to very-low-frequency electromagnetic energy. *Nav Res Rev* **1968**: 1–4, 1963.

Frey AH, Feld SR. Avoidance by rats of illumination with low power nonionizing electromagnetic energy. *J Comp Physiol Psychol* **89**: 183–8, 1975.

Frey AH, Feld SR, Frey B. Neural function and behaviour: defining the relationship. *Ann N Y Acad Sci* **247**: 433–439, 1975.

Fritze K, Sommer C, Schmitz B, Mies G, Hossmann KA, Kiessling M, Wiessner C. Effect of global system for mobile communication (GSM) microwave exposure on blood-brain barrier permeability in rat. *Acta Neuropathol* **94**: 465–470, 1997.

Fritze K, Wiessner C, Kuster N, Sommer C, Gass P, Hermann DM, Kiessling M, Hossmann KA. Effect of global system for mobile communication microwave exposure on the genomic response of the rat brain. *Neuroscience* **81**: 627–639, 1997.

Garaj-Vrhovac V, Horvat D, Koren z. The relationship between colony-forming ability, chromosome aberrations and incidence of micronuclei in V79 Chinese hamster cells exposed to microwave radiation. *Mutat Res* **263**: 143-149, 1991.

Garaj-Vrhovac V, Fucic A, Horvat D. The correlation between the frequency of micronuclei and specific chromosome aberration in human lymphocytes exposed to microwaves. *Mutat Res* **281**: 181–186, 1992.

Garaj-Vrhovac V, Fucic C, Pevalec-Kozlina B. The rate of elimination of chromosomal aberration after accidental exposure to microwaves. *Bioelectrochem Bioenerg* **30**: 315–325, 1993.

Garaj-Vrhovac V. Micronucleus assay and lymphocyte mititic activity in risk assessment of occupational exposure to microwave radiation. *Chemosphere* **39**: 2301-2312, 1999.

Garland FC, Shaw E, Gohan ED, Garland CF, White MR, Sinsheimer PJ. Incidence of leukemia in occupations with potential electromagnetic field exposure in United States Navy Personnel. *Am. J. Epidermiol.* **132**: 293–303, 1990.

Goldoni J. Hematological changes in peripheral blood of workers occupationally exposed to microwave radiation. *Health Physics* **58**: 205–207, 1990.

Goldsmith JR. Epidermiological evidence of radiofrequency radiation (microwave) effects on health in military, broadcasting and occupational studies. *Int J Occop Environ Health* **1**: 47–57, 1995.

Goldsmith JR. Epidermiological evidence relevant to radar (microwave) effect. *Environ. Health Perspect. 105 Suppl* **6**: 1579–1587, 1997.

Gorbach IN. changes in nervous system of individuals exposed to microradiowaves for long period of time. *Zdravookhraneniye belorussii* **5**: 51–53, JPRS 818654, pp. 24–28, 1982.

Goud SN, Usha Rani MV, Reddy PP, Reddi OS, Rao MS, Saxena VK. Genetic effects of microwave radiation in mice. *Mutat Res* **103**: 39–42, 1982.

Grajewski B, Cox C, Schrader SM, Murray WE, Edwards M, Turner TW, Smith JM, Shekhar SS, Evenson DP, Simon SD, Conover DL. Semen quality y and hormone levels among radiofrequency heat operators. *J Occup Envirn Med* **42**: 993–1005, 2000.

Grayson JK. Radiation exposure, socio-economic status and brain tumor risk in the U.S. Air Force: a nested case control study. *Am J Epidermiol* **143:** 480–486, 1996.

Grin AN. Effects of microwave on catecholamine metabolism in brain. US Joint Pub. Research Device Rep. JPRS 72606, 1974.

Groves FD, Page WF, Gridley G. Lisimaque L, Stewart PA, Tarone RE, Gail MH, Boice JD Jr, Beebe GW. Cancer in Korean war navy technicians: mortality survey after 40 years. *Am J Epidemiol* **155:** 810–818, 2002.

Haider T, Knasmueller S, Kundi M, Haider M. Clastogenic effects of radiofrequency radiations on chromosomes of Tradescantia. *Mutat Res* **324:** 65–68, 1994.

Hall AS, Prior MV, Hand JW, Young IR, Dickinson RJ. Observation by MR imaging of *in vivo* temperature changes induced by radiofrequency hyperthermia. *J Comput Assis Tomogr* **3:** 430–436, 1990.

Hanson Mild K, Oftedal G, Sandstrom M, Wilen J, Tynes T, Haugsdal B, Hauger E. Comparison of symptoms experienced by users of analogue and digital mobile phones: a Swedish-Norwegian epidemiological study. *Arbetslivsrapport* **23,** 1998.

Hardell L, Nasman A, Pahlson A, Hallquist A, Hansson Mild K. Use of cellular telephones and the risk for brain tumours: A case-control study. *Int J Oncol* **15:** 113–116, 1999.

Hayes RB, Morris Brown L, Pottern LM, Gomez M, Kardaun JWPF, Hoover RN, O'Connell KJ, Sutzman RE, Javadpour N. Ocupational and risk for testicular cancer: A case-control study. *Int J Epidemiol* **19:** 825-831, 1990.

Hocking B. Gordon I, Grain JL, Hatfield GE. Cancer incidence and mortality and proximity to TV towers. *Med J Aust Assoc* **165:** 601–605, 1996.

IEEE Standard for safety levels with respect to human exposure to radiofrequency electromagnetic fields 3kHz-300GHz. (Piscataway NJ: IEEE), C95. 1–1999.

INIRC-IRPA (International Non Ionizing Radiation Committee-International Radiation Protection Association) Guidelines on limits of exposure to radiofrequency fields in the frequency range from 100 kHz to 300GHz. *Health Phys* **54:** 115–123, 1988.

Inaba R, Shisido K, Okada A, Moroji T. Effects of whole body microwave exposure on the rat brain contents of biogenic amines. *Eur Appl Physiol* **65:** 124–128, 1992.

Jauchem JR, Frei MR. Cardiorespiratory changes during microwave-induced lethal heat stress and beta-adrenergic blockade. *J Appl Physiol* **77:** 434–40, 1994.

Jauchem JR, Ryan KL, Frei Mr. Cardiovascular and thermal responses in rats during 94 GHz irradiation. *Bioelectromagnetics* **20:** 264–7, 1999.

Johnson RB, Spackman D, Crowley J, Thompson D. Chou CK, Kunj LL Guy AW. Open field behavior and corticosterone. Effects of Long-term Low-level Radiofrequency Radiation Exposure on Rats. Brooks Air Force Base, Texas, USAF School of Aerospace Medicine Volume 4, USAFSAM TR 33-43, 1983.

Jeffreys A J, Wilson V, Thein S L. hypervariable "minisatellite" regions in human DNA. *Nature* **316:** 67–73, 1985.

Kalns J, Ryan KL, Mason PA, Bruno JG, Gooden R, Kiel JL. Oxidative stress precedes circulatory failure induced by 35-GHz microwave heating shock **13:** 52-59, 2000.

Kellenyi L, Thuroczy G, Faludy B, Lenard L. Effects of mobile GSM radiotelephone exposure on the auditory brainstem response (ABR). *Neurobiology* **7:** 79–81, 1999.

Kerbacher, JJ, Meltz ML, Erwin DN. Influence of radiofrequency radiation on chromosome aberrations in CHO cells and its interaction with DNA-damaging agents. *Radiat Res* **123:** 311–319, 1990.

Khalil AM, Qassem WF, Suleiman MM. A preliminary study on the radiofrequency field-induced cytogenetic effects in cultured human lymphocytes. *Dirasat* **20:** 121–130, 1993.

Klimkova-Deutschova E. Neurologic findings in persons exposed to microwaves. *Biologic Effects and Health Hazards of Microwave Radiation, Proceedings of an International Symposium*, Warsaw, 15-18 Oct. 1973, P. Czerski et al., eds., pp. 268–272, 1974.

Koenig H, Goldstone AD, Lu CY, Trout JJ. Polyamines and Ca^{2+} mediate hypersmolal opening of the blood-brain barrier: In vitro studies in isolated rat cerebral capillaries. *J Neurochem* **0:** 1135–1142, 1989.

Krause CM, Sillanmaki L, Koivisto M, Haggqvist A, Saarela C, Revonsuo A, Laine M, Hamalainen H, Effects of electromagnetic field emitted by cellular phones on the EEG during a memory task. *Neuroreport* **11(4):** 761–764, 2000.

Kwee S, Raskmark P. Changes in cell proliferation due to environmental non-ionizing radiation 2. microwave radiation. *Bioelectrochem Bioenerg* **44:** 251–255, 1998.

Lai H, Horita A, Chou CK, Guy AW. Effects of low-level microwaves irradiation on hippocampal and frontal cortical choline uptake are classically conditionable. Pharmocol. *Biochem Behav* **27:** 635–639, 1987a.

Lai H, Hortia A, Guy AW. Acute low-level microwave exposure and central cholinergic activity: studies on irradiation parameters. *Bioelectromagnetics* **9:** 355–362, 1988.

Lai H, Carino MA, Horita A, Guy AW. Low-level microwave irradiation and central cholinergic systems. Pharmacol. *Biochem Behav* **33:** 131–138, 1989a.

Lai H, Carino MA, Horita A, Guy AW. Low-level microwave irradiation and central cholinergic activity: a dose-response study. *Bioelectromagnetics* **10:** 203–208, 1989b.

Lai H. Research on the neurological effects of non-ionizing radiation at the University of Washington. *Bioelectromagnetics* **13:** 513–526, 1992.

Lai H, Carino MA, Horita A, Guy AW. Opioid receptor subtypes that mediate a microwaves-induced decrease in central cholinergic activity in the rat. *Bioelectromagnetics* **13:** 237–246, 1992.

Lai H, Horita A, Guy AW. Microwave irradiation affects radial-arm maze performance in the rat. *Bioelectromangetics* **15:** 95–104, 1994.

Lai H, Singh NP. Acute low-intensity microwave exposure increases DNA single-strand breaks in rat brain cells. *Bioelectromagnetics* **16:** 207–210, 1995.

Lai H, Singh NP. Single-and double-strand DNA braks in rat brain cells after acute exposure to radiofrequency electromagnetic radiation. *Int J Radiat Biol* **69:** 513–521, 1996.

Lai, H, Carino, MA, Singh, NP. Naltrexone blocks RFR-induced DNA double strand breaks in rat brain cells. *Wireless networks* **3:** 471–476, 1997.

Lai H. Neurological effects of radiofrequency electromagnetic radiation. "Workshop on Possible Biological and Health Effects of RF Electromagnetic Fields", Mobile Phone and Health Symposium, Oct 25–28, 1998, University of Vienna, Vienna, Austria. Published on www.mapcruzin.com

Lancranjan I, Maicanescu M, Rafaila E, Klepsch I, Popescu HI. Gonadic function in workmen with long-term exposure to microwaves. *Health Physics* **29:** 381–383, 1974.

Liakouris AGJ. Radiofreqency (RF) sickness in the Lilienfeld Study: an effect of modulated microwaves? *Arch Environ Health* **53:** 236-238, 1998.

Lilienfeld AM, Tonascia J, Tonascia S, Libauer CA, Cauthen GM. Foreign Service Health Status Study: Evaluation of Health Status of Foreign Service and Other Employees from Selected Eastern European Posts. Final Report Contract 6025-619073 (NTIS OB-288163). Washington: U.S. Department of State, 1978.

Lin RS, Dischinger PC, Conde J, Farrell KP. Occupational exposure to electromagnetic fields and occurrence of brain tumors. *J Occupat Med* **27**: 413–419, 1985.

Lin, J.C. Biological Effects Basis for Exposure Limits Panel (Invited), Radiofrequency Radiation Conference, U.S. Environmental Protection Agency, Bethesda, MD, 1993.

Lagorio S, Rossi S, Vecchia P, De Santis M, Bastianini L, Fusilli M, Ferrucci A, Desideri E, Comba P, Mortality of plastic-ware workers exposed to radiofrequencies. *Bioelectromagnetics* **18**: 418-421, 1997.

Lotz WG. Hyperthermia in radiofrequency-exposed rhesus monkeys: a compariso of frequency and orientation effects. *Radiat Res* **102**: 59–70, 1985.

Lu ST, Mathur SP, Akyel Y, Lee JC. Ultrawide-band electromangetic pulses induced hypotension in rats. *Physiol Behav* **65**: 753–761, 1999; Corrected and republished in Physiol Behav **67**: 753–761, 1999.

Lukashevsky KV, Belyaev IY. Switching of prophage gene in E. coli by millimeterwave. *Meas Sci Res* **18**: 955–957, 1989.

Maes A, Collier M, Slaets D, Verschaeve L. Cytogenetic effects of microwaves from mobile communication frequencies (954 MHz). *Electro Magnetobiol* **14**: 91–98, 1995.

Maes A, Collier M, Van Gorp U, Vandoninck S, Verschaeve l. Cytogenetic effects of 9352-MHz (GSM) microwaves alone and in combination with mitomycin. *C Mut Res* **393**: 151–156, 1997.

Maes A, Verschaeve L, Arroyo A, de Wagter C, Vercruyseen L. *In vitro* cytogenetic effects of 2450 MHz waves on human peripheral blood lymphocytes. *Bioelectromagnet* **14**: 495–501, 1993.

Mann K, Roschke J. Effects of pulsed high frequency electromagnetic fields on human sleep. *Neuropsychobiology* **33**: 41–47, 1996.

Markarov G, et al. Hypersensitivity to EMF, and the dependence of brain bioelectrical activity and general hemodynamics in cerebral asthenic (CA) patients, exposed to radioactive irradiation upon EMF 20-80 Hz effect, *Proceedings of the 2nd Copenhagen Conference on Electromagnetic Hypersensitivity*, May 1995. Katajainen J, Knave B, eds., pp. 57-60, 1995.

Martens L, DeMoerloose J, DeWagter C, DeZutter D. Calculaion of the electromagnetic fields induced in the head of an operator of a cordless telephone. *Radio Sci* **30**: 415–420, 1995.

Mason PA, Walters TJ, DiGiovanni J, Beason CW, Jauchem JR, Dick EJ Jr, Mahajan K, Dusch SJ, Shields BA, Merritt JH, Murphy MR, Ryan KL. Lack of Effect of 94.0 GHz Radio Frequency Radiation Exposure in an Animal Model of skin Carcinogenesis. Submitted for publication, 2001.

Malayapa RS, Ahern EW, Bi C, Straube WL, LaRegina M, Pickard WF, Roti-Roti JL Measurement of DNA damage in rat brain cells after in vivo exposure to 2450 MHz electromangetic radiation and various methods of euthanasia. *Radiation Research* **149**: 637–645, 1998.

McRee DI, Wachtel H. Pulse microwave effects on nerve vitality. *Radiat Res* **91(1)**: 212–218, 1982.

Meltz ML, Walker KA, Erwin DN. Radiofrequency (microwave) radiation expsoure of mammalian cells during UV-induced DNA repair synthesis. *Radiat Res* **110:** 255–266, 1987.

Meltz ML, Eagan P, Ersin DN. Absence of mutagenic interaction between microwaves and mitomycin C in Mammalian cells. *Environ Mol Mutagen* **13:**294–303, 1989.

Meltz ML, Holahan PK, Smith ST, Kerbacher JJ, Ciaravino V. Interaction of ionizing radiation, genetically active chemicals, and readiofrequency radiation in human and rodent cells. Department of Radiology. University of Taxas Health Science Center. USAF-SAM-TR-90-18, 1990.

Merritt JH, Shelton WW, Chamness AF. Attempts to alter $^{45}Ca^{2+}$ binding to brain tissue with pulsemodulated microwave energy. *Bioelectromagnetics* **3:** 475–478, 1982.

Michelozzi P, Ancona C, Fusco D, Forastiere F, Perucci CA. Risk of leukemia and residence near a radio transmitter in Italy. *Epidemiology* **9** (Suppl): 354, 1998.

Mickley GA, Cobb BL, Mason PA, Farrell S. Disruption of a putative working memory task and selective expression of brain c-fos following microwave-induced hyperthermia. *Phys Behav* **55:** 1029–1038, 1994.

Milham S Jr. Mortality in workers exposed to electromagnetic fields. *Environ Health Perspectives* **62:** 297–300, 1985.

Miller MW, Ziskin MC. Biological consequences of hyperthermia. *Ultrasound Med Biol* **15:** 707-722, 1989.

Modak AT. Stavinoha WB, Dean UP. Effect of short electromangetic pulses on brain acetylcholine content and spontaneous motor activity in mice. *Bioelectromagnetics* **2:** 89–92, 1981.

Moriyama E, Salcman M, Broadwell RD. Blood-brain barrier alteration after microwave-induced hyperthermia is purely a thermal effect. 1: temperaur and power measurements. *Surg Neurol* **35:** 177–182, 1991.

Muhm JM. MOrtality investigation of workers in an electromangetic pulse test program. *J Occup Med* **34:** 287–292, 1992.

Neubauer C, Phelan AM, KUes H, Lange DG. Microwave irradiation of rats at 2.45 GHz activates pinocytotic-like uptake of tracer by capillary endothelial cells of cerebral cortex. *Bioelectromangetics* **11:** 261–268, 1990.

NCRP. Biological Effects and Exposure Criteria for Radiofrequency Electromagnetic Fields, Report No. 86. National Council on Radiation Protection and Measurement, Bethesda, Maryland, 1986.

NRPB. ICNIRP guidelines for limiting exposure to time-varying electric, magnetic and electromangetic fields (up to 300 GHz): advice on aspects of implementation in the UK. Doc NRPB, **10:** 5, 1999.

Oscar KJ Hawkins TD. Microwave alteration of the blood-brain barrier system of rats. *Brain Res* **126:** 281–293, 1977.

Ouellet-Hellstrom Stewart WF. Miscarriages among female physiotherapists who report using radio-and microwave frequency electromangetic radiation. *Am J Epidemiol* **138:** 775–786, 1993.

Pakhomov AG Murphy MR. Low-Intensity Millimeter Waves as a Novel Therapeutic Modality. *IEEE Transactions on Plasma Science* **28:** 34-40, 2000.

Pakhomov AG, Akyel Y, Pakhomova ON, Stuck BE, Murphy MR. Current state and implications of research on biological effects of millimeter waves: a review of the literature. *Bioelectromanetics* **19:** 393, 1998.

Pasche B, Erman M, Hayduk R, Mitler MM, Reite M, Higgs L, Kuster N, Rossel C, Dafni U, Amato D, Barbault A, Lebet JP. Effects of low energy emission therapy in chronic psychophysiological insomnia. *Sleep* **19**: 327–336, 1996.

Paulraj R, Behari J. Radio frequency radiation effects on protein kinase C activity in rats' brain, *Mutat Res* **545**: 127–130, 2004.

Phelan AM, Lange DG, Kues HA, Lutty, Ga. Modification of membrane fluidity in melanin containing cells by low level microwave radiation. *Bioelectromagnetics* **13**: 131–146, 1992.

Ramel C. The nature of spontaneous mutations. *Mutation Research* **212**: 199–202, 1989.

Repacholi MH. Low Level Exposure to Radiofrequency Electromagnetic Fields: Health Effects and Research Needs. *Bioelectromagnetics* **19**: 1-19, 1998.

Roberts NJ Jr., Michaelson SM, Lu ST. The biological effects of radiofrequency radiation: a critical review and recommendations. *Int J Radiat Biol* **50**: 379-420, 1986.

Romano-Spica V, Muccin N, Ursini CL, Ianni A, Bhat NK. *Sts*1 oncogene induction by ELF-modulated 50 MHz radiofrequency electromangetic field. *Bioelectromangetics* **21**: 8–18, 2000.

Reiser H, Dimpfel W, Schober F. The influence of electromangetic fields on human brain activity. *Eur J. Med Res* **1**: 27–32, 1995.

Riu PJ, Foster KR, Blick DW, Adair ERA. Thermal Model for Human Thresholds of Microwave-Evoked Warmth Sensations. *Bioelectromagnetics* **18**: 578–583, 1997.

Robinette CD, Silverman C, Jablon S. Effects upon health of occupational exposure to microwave radiation (radar). *Am J Epidemiol* **112**: 39-53, 1980.

Rojavin MA, Ziskin MC. Medical Application of Millimeter Waves. *Q J Med* **91**: 57–66, 1998.

Rojavin MA, Ziskin MC. Effect of millimeter waves on survival of UVC-exposed Escherichia coli. *Bioelectromangetics* **16**: 188, 1995.

Rothman KJ, Chou CK, Funch DP, Dreyer NA. Assessment of cellular telephone and other radiofrequency exposers for epidemiological research. *Epidemiology* **7**: 291-298, 1996.

RSC. EPR 99-1. A Review of the Potential Health Risks of Radiofrequency Fields from Wireless Telecommunication Devices. An Exprt Panel Report prepared at the request of the Royal Society of Canada for Health Canada, 1999.

Ryan KL, D'Andrea JA, Jauchem JR, Mason PA. Radio Frequency Radiation of Millimeter Wave Length: Potential Occupational Safety Issues Relating to Surface Heating. *Health Physics* **78**: 170–181, 2000.

Sadchikova MN. State of the nervous system under the influence of UHF. *The Biological Action of Ultrahigh Frequencies*, Letavet AA, Gordon ZV, eds., Academy of Medical Sciences, Moscow, pp. 25-29, 1960.

Sadchikova MN. Clinical manifestations of reactions tomicrowave radiation in various occupational groups. *Biologic Effects and Health Hazards of Microwave Radiation, Proceedings of an International Symposium*, Warsaw, 15-18 Oct., P. Czerski et al., eds., pp. 261–267, 1973.

Sadchikova MN, et al., Significance of blood lipid and electrolyte disturbances in the development of some reactions to microwaves. *Gigiyena Truda i Professional'nyye Zabolevaniya* **2**: 38–39, JPRS 77393, pp. 37–39, 1980.

Salford LG, Brun A, Eberhardt JL, Malmgren L, Persson BRR. Electromagnetic field-induced permeability of the blood-brain barrier shown by immunohistochemical methods. *In : Resonance Phenomena in Biology,* ed. B Norden, C Ramel, Oxford: Oxford University Press : 87–91, 1992.

Salford LG, Brun A, Sturesson K, Eberhardt JL, Persson BRR. Permeability of the blood-brain barrier induced by 915 MHz electromanetic radiation, continuous wave and modulated at 8, 16, 50, 200 Hz. *Microscopy Research and Technique* **27**: 535–542, 1994.

Sanders AP, Joines WT, Allis JW. Effect of continuous-wave, pulsed, and sinusoidal-amplitude-modulated microwaves in brain energy metabolism. *Bioelectromagnetics* **6**: 89–97, 1985.

Sandstrom M, et al., Skin symptoms among VDT workers related to electromagnetic fields-a case referent study. *Indoor Air* **5**: 29-37, 1995.

Sarkar S, Ali S, Behari J. Effect of low power microwave on the mouse genome: A direct DNA analysis. *Mutat Res* **320**: 141–147, 1994.

Sarkar S, Ali S, Thelma BK, Behari J. Study of the mutagenic potential of low power microwaves by direct DNA analysis. Proc. Internatioanl Conference on Radiation Protection, Vienna, Austria, Vol 3, pp 565-567, 1996.

Sarkar S, Gupta MM, Selvamurthy W. Biological consequences of microwave stress. Implication for mutagenesis and carcinogensis. *IETE Tech Rev* **14**: 153–163, 1997.

Sarkar S. Ali S. Tandem repeat sequences as markers to study microwave DNA interaction. Proc. Symp. Low Level Electromagnetic field phenomenon in Biological Systems, India, pp. 80–83, 1999.

Sarkar S, Selvamurthy W. Radiation Hazards and Issues in Health Effects. All India Conference on EMI-EMC Issues (Classified), Bangalore July 19–21, 2001.

Sarkar S. Radiofrequency Radiation Effect in Human Lymphocytes. All India Conference on EMI-EMC Issues (Classified) Bangalore July 19–21, 2001a.

Sarkar S. Studies on Mutagenic and Carcinogenic Potential of Microwaves, if any. Closure Report No. DIPAs/10/2001, (2001b).

Saunders RD, Darby SC, Kowalczuk CI. Dominant Lethal studies in male mice after exposure to 2450 MHz microwave radiation. *Mutat Res* **117**: 345-356, 1983.

Saunders RD, Kowalczuk CI, Beechey CV, Dunford R. Studies in the induction of dominant lethals and translocations in male mice after chronic exposure to microwave radiation. *Int J Radiat Biol* **53**: 983–992, 1998.

Schilling CJ. Effects of acute exposure to ultrahigh radiofrequency radiation on three antenna engineers. *Occup Environ Med* **54**: 281-284, 1997.

Schilling CJ. Effects of exposure to very high frequency radiofrequency radiation on six antenna engineers in two separate incidetns. *Occup Med* **60**: 49–56, 2000.

Schirmacher A, Bahr A Kullnick U, Stoegbauer F Electromangetric fields (1.75 GHz) influence the permeability of the blood-brain barrier in cell culture model. Presented at the Twentieth Annual Meeting of the Bioelectromangetics Society, St. Pete Beach, FL, June 1998.

Schrader SM, Langford RE, Turner TW, Breitenstein MJ, Clark JC, Jenkins BL, Lundyl DO, Simonl SD, Weyandtl TBI. Reproductive function in relation to duty assignments among military personnel. *Reprod Toxicol* **12**: 465–468, 1998.

Shandala MG, Dumanski UD, Rudnev MI, Ershova LK, Los IP Study of nonionizing microwave radiation effects upon the central nervous system and behaviour reaction. *Environ Health Perspect* **30**: 115–121, 1979.

Shelton WW Jr, Merritt JH. In vitro study of microwave effects on calcium efflux in rat brain tissue. *Bioelectromagnetics* **2**: 161–167, 1981.

Sheppard AR, Bawin SM, Adey WR. Models of long-range order in cerebral macromolecules: effect of sub-ELF and ofmodulated VHF and UHF fields. *Radio Sci* **14(6S)**: 141–145, 1979.

Snyder SH. The effect of microwave irradiation on the turnover rate of serotonin and norepinephrine and the effect of microwave metabolizing enzymes. Washington DC, US Army Medical Research and Development Command Final Report Contract No DADA 17-69-C-9144, 1971.

Stagg RB, Thomas WJ, Jones RA, Adey WR. DNA synthesis and cell proliferation in C6 glioma and primary glial cells exposed to 836.55 MHz modulated radiofrequency field. *Bioelectromagnetics* **18**: 230–236, 1997.

Stark KDC, Krebs T, Altpeter E, Manz B, Griot C, Abelin T. Absence of chronic effect of exposure to shortwave radio broadcast signal on salivary melatonin concentrations in dairy cattle. *J Pineal Res* **22**: 171-176, 1997.

Steneck NH, Cook HJ, Vander AJ, et al. The origins of U.S. safety standards for microwave radiation. *Science* **6**: 1230–37, 1980.

Stevens RG. lectric power use and breast cancer: a hypothesis. *Am J Epidemiol* **125**: 556–561, 1987.

Summary of an advisory panel. Health Council of Netherlands. Radiofrequency electromangetic field (300 Hz-300 GHz). *Health Physics* **75**: 51–55, 1998.

Sykes PJ, 1 Brett D. McCallum BD, Bangay MJ, Hooker AM, Morley AA. Effect of Exposure to 900 MHz Radiofrequency Radiation on Intrachromosomal Recombination in pKZ1 Mice. *Radiate Res* **156**: 495–502, 2001.

Szmigielski S. Cancer morbidity in subjects occupationally exposed to high frequency (radiofrequency and microwave) electromagnetic radiation. *Sci Total Environ* **180**: 9–17, 1996.

Szmigielski, S. Bortkiewicz A, Gadzicka E, Zmyslony M. Kubacki R, Alteration of diurnal rhythms of blood pressure and heart rate to workers exposed to radiofrequency electromangetic fields. *Blood Press Monit* **3**: 323–330, 1998.

Takashima S, Onaral B, Schwan HP. Effects of modulated RF energy on the EEG of mammalian brains: Effects of Acute and Chronic Irradiations. *Radiat Environ Biophys* **16(1)**: 15–27, 1979.

Thomas TL, Stolley PD, Stemhagen A, Fontham ETH, Bleecker ML, Stewart PA, Hoover RN. Brain tumour mortality risk among men with electrical and electronics jobs: A case-control study. *I Natl. Cancer Inst.* **79(2)**: 233–238, 1987.

Thuroczy G, Kubinyi G, Bodo M, Bakos J, Szabo LD. Simultaneous response of brain electrical activity (EEG) and cerebral circulation (REG) to microwave exposure in rats. Rev. *Environm Health* **10**: 135–148, 1994.

Tice RR, Hook GG, Donner M, McRee DI, Guy AW. Genotoxicity of radiofrequency signals. I. Investigation of DNA damage and micronuclei induction in cultured human blood cells. *Bioelectromangetics* **23**: 113–126, 2002.

Trinos MS. Frequency of diseases of digestive organs in people working under conditions of combined effect of lead and SHF-range electromagnetic energy. *Gigiyena i sanitariya*, no. 9: 93–94, JPRS 84221, pp. 23–26, 1982.

Tynes T, Hannevik M, Andersen A, Vistnes AI, Haldorsen T. Incidence of breast cancer in Norwegian female radio and telegraph operators. *Cancer Causes Control* **7(2)**: 197–204, 1996.

US FDA guidelines (Food & Drug Administration: *Guidance for Magnetic Resonance Diagnostic Devices-Criteria for Significant Risk Investigations*; ttp://www.fda.gov/ gov/cdrh/ode/magdev.html, 1997.

UNEP/WHO/IRPA, Electromagnetic Fields (300 Hz – 300 GHz). Environmental Health Criteria 137. Geneva, World Health Organization, 1993.

Varma MM, Dage EL, Joshi SR. Mutagenicity induced by non-ionizing radiation in swiss male mice, in: C.C. Johnson, M. Shore (Eds.). Biological Effects of Electromagnetic Waves. Vol. 1. Food and Drug Administration. US (FDA), Washington, DC. USNC/ URSI Annual Meeting—selected papers, Oct. 20–23, Boulder, CO, pp. 397–405, 1976.

Varma MM, Traboulay EA Jr. Evaluation of dominant lethal test and DNA studies in measuring mutagenicity caused by non-ionizing radiation. In Johnson CC, Shore M (eds): "Biological Effects of Electromagnetic Waves." Washington, DC: US Food and Drug Administration (FDA), (USNC/URSI Annual Meeting-Selected Paprs, Oct. 20-23, 1975, *Boulder, CO* 1: 386–396, 1976.

Van Leeuwen GM, Lagendijk JJ, Van Leersum BJ, Zwamborn AP, Hornsleth SN, Kotte AN, Calculation of change in brain temperatures due to exposure to a mobile phone. *Phys Med Biol* 44: 2367–2379, 1999.

Velizarov S, Raskmark P, Kwee, S. The effects of radiofrequency fields on cell proliferation are non-thermal. *Bioelectrochem Bioelectrochem Bioenerg* 48: 177–180, 1999.

Von Klitzing I. Low frequency pulsed electromagnetic fields influence EEG of man. *Phys Medica* 11: 77–80, 1995.

Vollrath L, Spessert R, Kratzsch T, Keiner M, Hollmann H. No short-term effects of high- frequency electromangetic fields on the mammalian pineal gland. *Bioelectromagnetics* 18: 376–387, 1997.

Walters TJ, Blick DW, Johnson Lr, Adair ER, Foster KR. Heating and Pain Sensation Produced in Human Skin by Millimeter Waves: Comparison to a Simple Thermal Model. *Health Physics* 78: 259–267, 2000.

Wang B, Lai H. Acute Exposure to Pulsed 2450 MHz Microwaves affects Water-Maze Performance of Rats. *Bioelectromagnetics* 21: 52-56, 2000.

Weyandt TB, Schrader SM, Turner TW, Simon SD. Semen analysis of military personnel associated with military duty assignments. *Reprod Toxicol* 10: 5221–528, 1996.

WHO. "Environmental Health Criteria 137: Electromagnetic Fields (300 Hz to 300 GHz)." Geneva; World Health Organization, pp. 80–180 and 290, 1993.

Wu Y, Jia Y, Guo Y, Zheng Z, Influence of EMP on the nervous system of rats. *ACTA Biophysica Sinica* 15: 152-157, 1999.

Yao KTS, Jiles MM. Effects of 2450 MHz microwave radiation on cultivated rat kangaroo cells. Biological Effects and Health Implications of Microwave Radiation (S F Cleary, Ed), Proceedings of Medical College of Virginia Symposium, 1969. Richmond VA, Department of Health, Education and Welfare, Public Health Service, Environmental Health Service, Bureau of Radiology, p. 123–133, 1970.

Zalyubovskaya NP, Kiselev RI. Effect of radio waves of a millimeter frequency range on the body of man and animals. *Gigiyena i Sanitariya* 8: 35–39, JPRS 72956, pp. 9– 15, 1978.

Ziskin MC. Medical aspects of radiofrequency radiation overexposure. *Health Physics* 82: 387–391, 2002.

Topics in Electromagnetic Waves: Devices, Effects and Applications
Edited by J. Behari
Copyright © 2005, Anamaya Publishers, New Delhi, India

8. Study of DNA Damage in Workers Occupationally Exposed to Xerox Machine

Y.R. Ahuja[1], K.I. Goud[2], Ravindra K. Tiwari[3] and S.C. Bhargava[4]

[1]Vasavi Medical & Research Centre, 6-1-91 Khairatabad, Hyderabad-500 004 and
Stem Cell and Molecular Biology Laboratory, Nizam's Institute of Medical Sciences,
Hyderabad-500 082, India

[2]Department of Molecular Diagnostics, Apollo Hospital, New Delhi-110 044, India

[3]Department of Zoology, Bhavan's New Science College, Hyderabad-500 029, India

[4]Department of Electrical & Electronic Engineering, MG Institute of Technology,
Gandipet, Hyderabad-500 075, India

Abstract: Xerox machine operation has become a common profession in urban
India. Every xerox machine invariably contains various electrical devices, which
are sources of extremely low frequency and possibly some higher frequencies of
electromagnetic fields (EMFs). In addition, the toner (ink) used in xerox machines
has colorants, polycyclic hydrocarbons and styrene. During running of the machines,
ozone, nitrogen oxide and volatile organic compounds are released. Operators are
exposed to a complex mixture of EMFs and a variety of chemicals. To evaluate the
genotoxic potential of exposure in xerox machine operators, comet assay on the
peripheral blood leucocytes was performed. In our study 202 exposed and 198 control
subjects were included. There was a significant increase in DNA damage in the
exposed as compared to the control subjects. Of the various confounding factors
analysed, duration of work and smoking had a positive correlation. An increase in
DNA damage may have long range health effects like infertility, congenital
malformations and cancer.

Introduction

With the advent of industrialization our exposure to electromagnetic fields
(EMFs) has increased enormously in our day-to-day life. A large number of
epidemiological and laboratory studies have been carried out to assess the effects
of EMFs on health. However, the results are controversial [1]. Some studies
have suggested that EMFs by themselves may not have deleterious effects, but
there is a possibility of their interaction with the chemical pollutants in the
environment [2].

Operation of xerox machines has become a common source of employment
in urban India. The commercial market is flooded with different brands and
quality of such machines. Every xerox machine invariably contains various
electrical devices/parts at varying voltages and operating frequencies, which
are sources of different electromagnetic radiations, including namely the

extremely low frequency (typically 50 Hz) electromagnetic field (EMF). In addition, toner (ink) used in xerox machines, to get a permanent image on paper, has three essential components: colorants (carbon black being the most common), polycyclic aromatic hydrocarbons (PAHs) and styrene. During running of the machine ozone (O_3), nitrogen oxide (NO) and volatile organic compounds (VOCs) are released. Due to proximity of the operator during running of the machine, exposure to extremely low and some higher frequency EMFs for 8-10 h per working day over a period of years appears to be substantial. At the same time, operators have an exposure to toners during the loading process and to a complex mixture of O_3, NO, VOCs, during the operation of the xerox machine [3].

To the best of our knowledge, no information is available on the genotoxic potential of exposure of the workers occupationally exposed to xerox machines. Therefore, we decided to conduct such a study. To evaluate genotoxic potential, alkaline comet assay was used. Comet assay is the most sensitive test known for the detection of DNA damage at the single cell level. It requires only a small cell sample ($\sim 10,000$ cells) and results can be obtained in relatively short time (6-8 h).

Material and Methods

A well designed proforma was completed in order to collect detailed information from the individuals on age, duration of work, health history, family history, diet and habits (smoking and drinking).

Our exposed population consisted of 202 males (age range of 17-40 years), working 8-10 h per day with photocopying machines, for more than a year. As controls, 198 sex and age matched individuals working in different professions (like clerks, attendents and students) and with the same socio-economic status were included. From each of the subjects 40 µl of peripheral blood was drawn by finger prick in a heparinized capillary tube. The blood samples were analyzed for DNA damage by alkaline comet assay after the method of [4]. Slides thus prepared were stained after the method of [5].

The slides were screened under a bright field transmission light microscope. Using an ocular micrometer comet tail length (Fig. 1), which is an estimate of DNA damage was measured in 100 cells per treatment and its mean ± SE calculated.

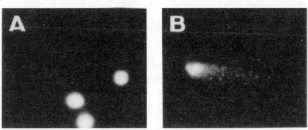

Fig. 1 (A) Undamaged cell appearing as a halo and (B) a cell with DNA damage appearing as a comet.

The data obtained from all samples, for each parameter, were pooled and mean ± SE was calculated for each group. Student's t-test was performed to evaluate various differences. Linear regression analysis was carried out to study the effects of confounding factors.

Results

There was a significant increase in mean comet tail length (DNA damage)in the exposed as compared to the control subjects (Table 1). Among confounding factors, duration of work and smoking showed a significant effect on DNA damage in the exposed subjects (Table 2).

Table 1.   Mean ± SE of comet tail length (DNA damage) in leucocytes of exposed and control subjects

Subjects	Sample size	No. of cells scored	Tail length (μ)
Exposed	202	20200	15.12 ± 0.29*
Control	198	19800	8.05 ± 0.14

*p<0.01

Table 2.   Effect of duration of work-exposure and smoking on DNA damage in exposed subjects (sample size is given in parenthesis).

Duration of work-exposure (years)	Tail length (μ)	Smoking habit	Tail length (μ)
1-5(112)	12.37 ± 0.24	NS (116)	7.72 ± 0.29*
6-10(90)	18.53 ± 0.33*	S(86)	11.59 ± 0.24

*p < 0.01, S: Smoker, NS: Non-smoker

Discussion

In the present study, the increased DNA damage in the exposed subjects may be due to a chronic exposure to complex mixture of extremely low and some higher frequency EMFs, components of tonners and emissions from photocopiers, as well as an interaction between them.

In our laboratory we have studied the effects of varying flux densities of ELF-EMFs (0.2, 0.4, 0.6, 0.8, 1.0, 2.0, 3.0, 5.0, 7.0 and 10 mT) at 50 Hz on human lymphocytes. The exposure was given for one hour in each case. With increase in magnetic field flux density there was a stepwise significant increase in DNA damage [6]. It is generally accepted that MFs do not generate enough energy to directly damage DNA. However, it has been suggested that MFs may have the potential of altering chemical processes in which free radicals, which may be responsible for causing DNA damage, are involved [7].

DNA damage is an indicator of genetic mutations/genomic instability (genetic as well as epigenetic changes). Mutations may be the cause of generic disease like mental retardation and congenital malformations, whereas genomic instability is usually associated with cancer. Since the latent period of cancer is long, it is difficult to see the outcome of studies using genomic instability as a biomarker. However, wherever possible, efforts are being made to follow the high risk individuals. For example, the Nordic study group has verified the association of the magnitude of chromosome damage, which is also a biomarker of genomic instability, with cancer risk [8].

There are a large number of studies on health risk from exposure to EMFs from gadgets used at home, office or hospital. Some of these studies will be mentioned here.

A side effect of industrialization has been environmental pollution. A large number of epidemiological and laboratory studies have been carried out to evaluate the health effects of these pollutants. These studies emphasize the growing impact of environmental chemical pollution and the rapidly increasing deployment of an almost infinite variety of environmental electromagnetic fields as possible joint factors in cancer promotion [2].

An increased risk of leukemia, brain tumor and prostate cancer has been reported in workers exposed to electrical and magnetic fields [9-10]. The occupations grouped as sharing exposure to electric and magnetic fields undoubtedly share other exposures like heat, metal fumes, solvents (including benzene). Polychlorinated biphenyls, synthetic waxes, epoxy resins and chlorinated naphthalenes.

Occupational exposure to xerox machines involves EMFs, toners and the fumes evolved during operation. Earlier studies have revealed that exposure to these compounds induce mutagenic/carcinogenic effects [11]. It is also probable that EMF exposure in xerox machine operators may have potentiated the effects of these compounds by facilitating their absorption [12].

There are a large number of reports on cancer-related biological effects of EMFs. Different animal models and study designs have been used to address possible carcinogenic effects of MFs. Based on a comparison of the results, it has been proposed that MF exposure may potentiate the effects of known carcinogens only when both exposures are chronic [13]. This, of course, would be relevant to human exposure, because exposures of humans to known carcinogenes are typically long term.

Conclusion

Our study on commercial xerox machine operators has shown a significant increase on DNA damage in their white blood cells as compared to the controls. An increase in DNA damage is an indicator of risk for genetic diseases including cancer, which may have long (10-15 years) gestation period. Precautionary measures for these operators are recommended.

Acknowledgements

Our thanks are due to Prof. K. Prabhakar Rao, Mr. J.D. Fernendes and Dr Q. Hasan for their suggestions, to Dr. K.V.R, Dr. N. Balakrishna for statistical advice and to Mr. K.J. Tulja Ram, Ms. P. Shivani and Mr. V. Sridhar Reddy for secretarial assistance.

References

1. A. Ahlbom, E. Cardis, A. Green, M. Linet, D. Savitz and A. Swerdlow, (ICNRP: International Commission for Non-Ionizing Radiation Protection), "Review of the epidemiologic literature on EMF and health", *Environ. Health Perspect.*, **109**, pp 911–934, 2001.

2. W.R. Adey, "Joint actions of environmental non-ionizing electromagnetic fields and chemical pollution in cancer promotion", *Environ. Health Perspect.* **86**, pp 297–305, 1990.

3. S.K. Brown "Assessment of pollutant emissions from dry photocopiers", *Indoor-Air*, **9**, pp 259–267, 1999.

4. N.P. Singh, M.T. McCoy, R.R. Tice and E.L. Schneider, "A simple technique for quantation of low levels of DNA damage in individual cells", *Exper. Cell Res.*, **175**, pp. 184–191, 1988.

5. Y.R. Ahuja and R. Saran "Alkaline single cell gel electrophoresis assay/Comet assay. I Protocol", *J. Cytol. Genet.*, **34**, pp 57–62, 1999.

6. R.K. Tiwari, K.I. Goud, K. Suryanarayana, S.C. Bhargava and Y.R. Ahuja. "DNA integrity in vitro study of human peripheral blood leucocytes exposed to ELF-EMFs using comet assay". Proceedings International conference on Electromagnetic Interference & Compatibility (INCEMIC), Bangalore, pp. 311–315, 2002.

7. H. Lai and N.P. Singh 'Melatonin and N-tert-butyl-a-phenylnitrone block 60 Hz magnetic field-induced DNA single and double strand breaks in rat brain cells". *J. Pineal Res.* **22**, pp. 152–162, 1997.

8. L. Hagmar, U.L.F. Stromberg. H. Tinnerberg and Z. Mikoczy. "The usefulness of cytogenetic markers as intermediate endpoints in carcinogenesis", Int. J. Hygiene Environ. *Health,* **204**, pp. 43-47, 2001.

9. M. Feychtung, U. Forssen and B. Floderus "Occupational and residential magnetic field exposure and leukemia and central nervous system tumors", *Epidemiology*, **8**, pp 384–389, 1997.

10. L.E. Charles, D. Loomis, C.M. Shy, B. Newman, R. Millikan, L.A. Nylander-French and D. Couper "Electromagnetic fields, polychlorinated biphenyls, and prostate cancer mortality in electric utility workers", *Am. J. Epidemiol*, **157**, pp 683–691, 2003.

11. G. Lofroth, E. Hefner, I. Alfthehm, and M.Moller. "Mutagenic activity in photocopiers". *Science* **209**, pp. 1037-1039, 1980.

12. L. Verschaeve and A. Maes "Genetic, carcinogenic and tetratogeneic effects of radiofrequency fields". *Mutat. Res.* **410**, pp. 141–146, 1995.

13. J. Juutilainen, S. Lang and T. Rytomaa. "Possible cocarcinogenic effects of ELF electromagnetic fields may require repeated long term interaction with known carcinogenic factors", *Bioelectromagnetics* **21**, pp. 122–128, 2000.

Topics in Electromagnetic Waves: Devices, Effects and Applications
Edited by J. Behari
Copyright © 2005, Anamaya Publishers, New Delhi, India

9. Electrical Characterization of Cataractous Lens

D.V. Rai, K.S. Kohli[1] and N. Goyal[2]

Department of Biophysics, Panjab University, Chandigarh-160014, India

[1]Department of Radiation Physics, Princess Margaret Hospital 610,
University Avenue, Toronto, ON M4G 2M9, Canada

[2]Centre of Advance Study of Physics, Panjab University,
Chandigarh-160 014, India

Abstract: Electrical properties of cataractous lens have been studied in the present work in the form of Cole-Cole plot. Twelve pairs of goat (*Caprus caprus*)'eye were taken within three to four hours after sacrificing animals. The lenses were removed from their capsules intact and divided into two groups, viz. control and experimental. The lens opacity *in vitro* was performed by standard methods. A sophisticated computer assisted AC impedance system (EG & G PARC Model 318) was used for measurement of AC impedance of control and cataractous lens over a frequency range from 10 mHz to 10 Hz. Two probe Ag-AgCl electrode systems were used for electrical measurements. The experiments were carried out at room temperature (24 ± 1°C).

Results have shown that the extracellular resistance Re and depressed angle θ approximated 48 ± 2.14 KΩ and $36 \pm 0.91°$ in control as compared to 38 ± 1.87 KΩ and $46 \pm 1.80°$ in the experimental group. The impedance locus of control lens shows a perfect arc of semi-circle while cataractous lens split in the plot. The frequency dependence of capacitance C, total impedance $|Z|$ and phase angle ϕ are also examined. An attempt has been made to explain the data in the light of compositional changes in the lens.

1. Introduction

Visual system is generally considered to be the most sensitive organ to non-ionising electromagnetic fields in the human body [1-4] and may also cause cataractogenesis [5]. Cataract is a clinical manifestation signifying the reduced visual performance due to light dissemination in the lens [6]. Increased light dissemination occurs as a result of an enhanced density of refractive index gradients in the lens. Vacuole formation and extracellular accumulation of water leads to regions of lowered refractive index [7]. Aggregation of lens proteins through disulfide bridges, non-disulfide covalent cross links and Ca^{2+} bridges [8], cause regions with increased refractive index. The loss of water of hydration in a protein molecule due to changed suprastructure, *synereis* [9] induces a reversible phase causing a refractive index gradient [10]. Membrane degeneration create disorder of the originally ordered array of lens membranes thereby increasing the intensity of light scattering due to less destructive interference [11]. The disorientation of cytoskeletal elements may cause fluctuations of optical anisotropy which increases the intensity of light scattered, resulted reduction of transmitted light.

Applied methods for cataract measurement include measurements of absorption properties [12, 13], image-forming properties [14], anatomical description [15], retroillumination photography [16], quasielastic light scattering [10] and computer-aided analysis of lenses imaged against a white background [17]. However, none of these clinical modalities provide direct evaluation of the integrity of extracellular, intracellular and cell-to-cell communication in cataract lens. Since electrical impedance technique is very precise approach in evaluating the structural compartment of the tissue [18, 19], it may be possible to see if cataract formation in the crystalline lens is related to the functional integrity of cell-to-cell junctions [20]. This hypothesis seems to be convincing in the light of data which suggest that in many forms of cataract, there is evidence of uncoupling of normally coupled cells [21, 22].

Lens opacification *in vitro* can be induced by various means [22-24]. The calcium-induced opacity is presently being used as a simulated model for cataract. The possibility that calcium has a role in cataractogenesis was recognised with the fact that calcium levels are high in many types of human cataracts [25]. Experimental studies on the calcium-induced cataract demonstrate that calcium concentration more than 3 mM can lead to the formation of cataract [24, 26, 27].

The aim of the present work is to study the electrical properties of normal and experimental cataractous lens in the form of Cole-Cole plot. The experiment was therefore designed to measure the complex impedance of normal and cataractous lens. The complex impedance plot was drawn between real Z' and imaginary Z'' parts as impedance locus between the frequency 10 mHZ to 10 Hz. Using graphical analysis of impedance locus, the extracellular resistance Re, depressed angle θ, distribution factor α and the frequency response of total impedance $|Z|$, capacitance C, phase angle ϕ were determined.

2.   Materials and Methods

2.1.   Sample Preparations

Twelve pairs of adult goat (*Caprus caprus*) eye were procured from the local abattoir, in a chilled hanks solution within three to four hours after killing of animals. The eyes were carefully enucleated and the eyeballs opened by a posterior approach. The lenses were cut free from the zonules and separated from the vitreous by sliding the adherent vitreous without touching the posterior surface. The lenses were removed with their capsules intact and divided into two groups, viz. control and experimental. The right lens of each pair was kept as control and left as experimental. In order to induce lens opacity *in vitro*, fresh lenses were incubated for 16 h at 37°C in 10 mM $CaCl_2$ solution made in tris-HCl buffer, 0.01 mM; pH 7.4 [27]. The control lenses were preserved in tris-

HCl buffer of 0.01 mM, pH 7.4. The normal and experimental cataractous lenses were placed on wire mesh to evaluate their optical properties.

2.2. Electrical Measurements

A sophisticated computer aided AC impedance system (EG & G, PARC, USA) was used for measurement of AC impedance [28] of normal and cataractous lens. Two probe Ag-AgCl electrode system [29] were used for electrical measurements. The electrical system is controlled with a computer with a dedicated software having facilities for automatic acquisition, merging and display of data. The software helps to measure the real Z' and imaginary Z'' parts of AC impedance at frequencies of operation. AC impedance system contains the complete hardware and software package, which includes a complex array of instruments and work together to provide excitation waveforms to the system and analyse its response. It includes electrometer, potentiostat galvanostat, two phase lock-in-amplifier and interface. An Epson Fx 80 printer with Graphics plus' was also used to make the system complete. The electrical measurements were performed at room temperature ($24 \pm 1°C$).

Fast Fourier Transform Technique (FFT) was used at lower frequencies with the help of potentiostat. The technique uses a computer algorithm to make the measurements in a significantly different way. A mixture of excitation waveforms was applied to the test system simultaneously. This complex waveform can be thought of as the algebraic sum of many individual waveforms of equal amplitude, each having a different frequency and phase characteristics. This kind of signal can be digitally generated and is called "pseudo random noise". Once the complex excitation waveform has been applied to the test system, an equally complex response is obtained. An inverse FFT operation is performed on this waveform to resolve it into discrete frequency data.

A major advantage of FFT technique is that low frequency data can be obtained faster than the conventional methods, thus minimising the polarisation times. For example, twenty frequencies between 0.001 to 0.1 Hz could be examined in a little over 100 s, while a frequency be frequency method would require this time to acquire 0.001 Hz data point alone.

Results

The electrical properties of normal and cataractous lens were recorded in the form of Cole-Cole plot. It shows a perfect arc of semicircle whereas cataractous lens shows a split in Cole-Cole plot. Analysis of Cole-Cole plot demonstrates in terms of various parameters in Table 1. The extracellular resistance Re of normal lens decreases (36 kΩ) as compared to the cataractous lens (46 kΩ). Similarly, the depressed angle for cataractous lens ($46 \pm 1°$) is higher than the normal lens ($36 \pm 0.91°$). The total impedance $|Z|$ and the phase angle ϕ of normal and cataractous lens are shown in Fig. 1. A reduction was found in absolute impedance and alteration in phase characteristics in cataractous lens as compared to normal lens.

Table 1.   Analysis of Cole-Cole plot of normal and cataractous lens

Lens	Depressed angle θ (degrees)	Extracellular resistance Re (kΩ)
Normal	36 ± 0.91	48 ± 2.14
Cataract	46 ± 1.0	38 ± 1.87

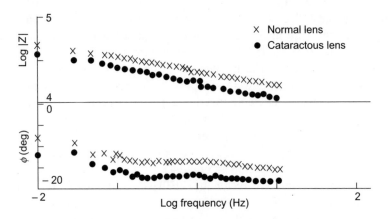

Fig. 1   Frequency response of total impedance $|Z|$ and phase ange φ.

Figure 2 shows the frequency response of capacitance, which was corrected for electrode polarization effects [28] for normal and cataractous lens. The higher value of capacitance is registered for cataractous lens as compared to the normal lens.

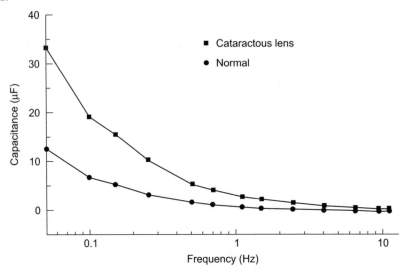

Fig. 2   Frequency response of capacitance.

Discussion

Investigations of lens opacity demand a suitable method for the differentiation between normal and pathological lenses. The electrical impedance technique is a precise and reliable modality in understanding the structural organisation and composition of the tissue [19, 30, 31, 32]. The flow of current in lens like other tissues is limited by highly insulating cell membranes. However, with rising alternating current frequency, the membrane impedes the flow of current. The measurements of electrical impedance over a frequency range allow separation of extracellular space and of the membranes themselves [20, 31, 33]. Electrical impedance technique allows gains in speed and simplicity over classical techniques such as dilution of radioactive ions for the measurement of extracellular resistance and membrane permeability [33].

The normal goat eye lens shows a perfect arc of semicircle with its centre lying below the abscissa. The control lens contains a singlet Cole-Cole plot. Interestingly, cataractous lens shows the presence of doublet Cole-Cole plot which reveals the presence of two phase system within the same lens. The present plot demonstrates the presence of normal and rapture membranes in the tissue. It may be suggested that the lower frequency Cole-Cole plot represents the nature of electrical flow in the region of tissue where the membranes are disrupted whereas the higher frequency plot represents the pathway in intact membrane. Also it may be assumed that lower frequencies can easily penetrate through the ruptured membranes, whereas these cannot pass through the intact membranes. On the other hand, higher frequencies can pass easily through the intact membranes [34]. It shows the two paths of current flows in the tissue.

The extracellular resistance in cataractous lens is considerably reduced as compared to normal lens. The reduction in extracellular resistance could be explained on the basis that cataractous lens has higher concentration of calcium [25, 27] which in turn helps in electrical conduction and thus reduces the extracellular resistance.

The inner membrane of the lens contribute to the electrical properties of the tissue, but have insulating properties more reminiscent of lipid bilayers (model membrane) than of biological membranes [35]. The inner membranes are excellent insulators allowing only a few ions per unit area to cross from cytoplasm to extracellular medium [36]. The total area of inner membranes is large. However, the total ionic current flow (across all the inner membranes) is almost equal to the ionic current which flows across the surface membranes. Indeed, if the inner membranes did not have such unusual insulating properties, almost all the conductance of the preparation would have been in its interior, in the combination of inner membranes and tortuous extracellular space [20]. The insulating properties of the inner membranes serve greatly to reduce the amount of sodium pumping and therefore metabolism needed to maintain the viability of the membranes permit substantial leakage of sodium ions and water into their cytoplasm and so most cells must expend substantial amounts of metabolic energy to maintain ionic and osmotic gradients [37]. If the inner membranes were not exceptionally good insulators, the lens would have needed a high density of sodium pumping activity, accompanying metabolism, and subcellular organelles to maintain its ionic and osmotic gradients. But even then, there would be serious diffusional difficulties introduced by the tortuous extracellular space and there might be optical difficulties as well, introduced by the subcellular organelles. Diffusion within the extracellular space could limit transport across the inner membrane, because the extracellular space within the lens is not a low resistance pathway. Even though the resistivity of the solution within the extracellular space is close to that of Ringer's solution, the total resistance of the extracellular space is very high comparable to the membrane resistance because of the small volume fraction and tortousity of the extracellular pathways [20]. Thus, ions transported across inner membranes would probably accumulate (or deplete) in unstirred layers within the extracellular medium. It is known that higher concentration of calcium in cataractous lens disrupt the membrane [27]. The disruptions of membranes may cause a reduction in capacitive impedance. The electrical measurement on cataractous lens shows a higher value of capacitance and thus reduction in capacitive impedance. The reduction in capacitive impedance has led to reduction in the overall impedance.

It may be concluded from the present study that the changes in various electrical parameters derived from Cole-Cole plot can be correlated with lens opacity. It is hoped that electrical modality may possibly be helpful in evaluating various lens disorders. The information on electrical changes that occur in lenses after incubation with high concentration of Ca^{2+} provides a basis for future investigations on the physiological understanding that are responsible for calcium-induced toxic cataract.

Acknowledgements

The authors would like to thank Mr Lalit M. Aggarwal for helping in data analysis and preparation of the manuscript.

References

1. Bernhardt JH. The establishment of frequency dependent limits for electric and magnetic fields and evaluation of indirect effects. *Radiat. Environ. Biophys.* 1988; **27:** 1–27.

2. Carstensen EL, Buettner A, Genberg VL and Miller MW. Sensitivity of the human eye to power frequency electric fields. *IEEE Trans. Biomed. Eng.* 1985; **32:** 561–5.

3. International Commission on Non-Ionizing Radiation Protection. Guidelines for limiting exposure to time-varying electric, magnetic, and electromagnetic fields (up to 300 GHz). *Health Phys.* 1998; **74:** 494–522.

4. Lindenblatt G and Silny J. A model of the electrical volume conductor in the region of the eye in the ELF range *Phys. Med. Biol.* 2001; **46:** 3051–3059.

5. Zaret, M.M. Electromagnetic energy and cataracts. In: A.A. Marino. Ed. Modern Bioelectricity. Marcel Dekker Inc., New York: 1988; 839–859.

6. Soderberg PG. Acute cataract in rat after exposure to radiation in the 300mm wavelength region, A study of the macro-micro-and ultrastructure. *Acta. Ophthalmol.* 1988; **66:** 141–52.

7. Philipson BT. Biophysical studies on normal and cataractous lenses. *Acta. Ophthalmol. Suppl.* 1969; 103.

8. Harding JJ, Dilley VJ. Structural proteins of the mammalian lens: a review with emphasis on changes in development, ageing and cataract *Exp. Eye Res.* 1976; **22:** 1–73.

9. Bettelheim FA. Syneresis and its possible role in cataractogenesis *Exp. Eye Res.* 1979; **28:** 189–97.

10. Tanaka T, Bendek GB. Observation of protein diffusivity in intact human and bovine lenses with application to cataract. *Invest. Ophthalmol.* 1975; **14:** 449–456.

11. Bettleheim FA, Chylack LT. Light scattering of whole excised human cataractous lenses. Relationships between different light scattering parameters. *Exp. Eye. Res.* 1985; **41:** 19–30.

12. Pirie A. Color and solubility of the proteins of human cataracts. *Invest. Ophthalmol.* 1968; **7:** 634–50.

13. Marcantonio JM, Duncan G, Davies PD, Bushell AR. Classification of human senile cataractous by nuclear colour and sodium content *Exp. Eye. Res.* 1980; **31:** 227–37.

14. Weale RA. Transparency and power of post-mortem human lenses. Variation with age and sex. *Exp. Eye. Res.* 1983; **36:** 731–41.

15. Chylack LT. Classification of human cataracts. *Arch. Ophthalmol.* 1978; **96:** 888–92.

16. Kawara T, Obazawa H. A new method for retroillumination photography of cataractous lens opacities *Am. J. Ophthamol.* 1980; **90:** 186–89.

17. Chylack T, Rosner B, White O, Tung W H, Sher LD. Standardisation and analysis of digitised photographic data in the longitudinal documentation of cataractous growth *Curr. Eye Res.* 1988; **7:** 223–35.

18. Ackmann JJ, Seitz MA. Methods of complex impedance measurements in biologic tissue CRC. *Crit. Rev. Bioeng.* 1984; **11:** 281–311.

19. Ackmann JJ. Complex bioelectric impedance measurement system for the frequency range of 5 Hz to 1 MHz. *Ann. Biomed. Eng.* 1993; **21:** 135–46.

20. Mathias RT, Rae JL, Eisenberg RS. Electrical properties of structural components of crystalline lens. *Biophys. J* 1979; **25:** 181–201.

21. Eisenberg RS, Rae JL. Current voltage relationships in the crystalline lens. *J. Physiol.* 1976; **262:** 285–300.

22. Harding J. Cataract: Biochemistry, Epidemeology and Pharmacology. Chapman and Hall, London 1991.

23. Ostadalova I, Babicky A, Obenberger J. Cataract induced by administration of a single dose of sodium selenite to suckling rats. *Experientia* 1978; **34:** 222–23.

24. Delamere NA, Paterson CA. Hypocalcaemic cataract. In: Mechanism of cataract formation in human lenses (Ed. G. Duncan) Academic Press, London, 1981.

25. Duncan G, Jacob TJC. The lens as a physicochemical system. In: The Eye. Vol. 1B (Ed. H. Davson) Academic Press, London, 1984.

26. Duncan G, Bushell AR. Ion analyses of human cataractous lenses *Exp. Eye. Res.* 1975; **20:** 223–30.

27. Clark JL, Mengel L, Bagg A, Benedek GB. Cortical opacity, calcium concentration and fiber membrane structure in calf Lens. *Exp. Eye Res.* 1980; **31:** 399–410.

28. Kohli KS, Rai DV, Kumar P, Jindal VK, Goyal N. Impedance of goat lens *Med. Biol. Eng. Comput.* 1997; **35:** 348–53.

29. Kohli KS, Rai DV, Jindal VK, Goyal N. Impedance of goat eye lens at different D.C. voltage. *Med. Biol. Engg. Comput.* 1998; **36(4):** 604–07.

30. Pethig R. Electrical properties of biological tissues. In: Modern Bioelctricity (Eds. AA Marino) Marcel and Dekker, New York, 1988.

31. Cornish BH, Thomas BJ, Ward LC. Improved prediction of extracellular and total body water using impedance loci generated by multiple frequency bioelectrical impedance analysis *Phys. Med. Biol* 1993; **38:** 337–46.

32. Ellis KJ. Human Body Composition: In Vivo Methods Physiol. Rev 2000; **80(2):** 649–80.

33. Heroux P, Bourdages M. monitoring living tissues by electrical impedance spectroscopy. *Ann. Biomed. Eng.* 1994; **22:** 328–37.

34. Cole KS. Membranes, Ions and Impulses. University of California Press, Berkeley and Los Angeles, 1972.

35. Alcala J, Massel H. Biochemistry of lens plasma membranes and cytoskeleton In: The ocular lens structure, function and pathology, Marcel and Dekker Inc., New York, 1985; 169–222.

36. Rae JL, Mathias RT. The physiology of the lens. In: The ocular lens: structure, function and pathology (Ed. H. Massel) Marcel Dekker, New York, 1985.

37. Tosteson DC, Hoffman JF. Regulation of cell volume by active cation transport in high and low potassium sheep red cells. *J. Gen. Physiol.* 1960, **44:** 169.

Topics in Electromagnetic Waves: Devices, Effects and Applications
Edited by J. Behari
Copyright © 2005, Anamaya Publishers, New Delhi, India

10. Therapeutic Effects of Millimeter Wave Resonance Therapy

Sushil Chandra, Geeta and Nikhilesh Kumar

Institute of Nuclear Medicine & Allied Sciences, New Delhi, India

Abstract: The therapeutic effects of low intensity millimeter wave resonance therapy (MRT) in frequency range of 53.33-78.33 GHz is studied. The reported success rate of millimeter wave resonance therapy for various pathologies is astonishingly high.

Introduction

The application of low-intensity mm waves in biomedicine is a new trend, originated in former USSR for the treatment of various diseases. Use of mm waves for medical purposes is known as "Microwave Resonance Therapy" (MRT). MMWR therapies and techniques comprise a part of a larger field known as bioelectromagnetism (BEM).

Millimeter wave belongs to relatively narrow range of electromagnetic waves with wavelengths from 1 to 10 mm. The part of mm waves used in therapeutics is 30-300 GHz. The penetration depth of millimetre wave into biological tissue is very small. A low average incident power density of 0.01 mW/cm^2, millimeter wave usually produce an average heating of an irradiated surface on the order of several tenths of a degree centigrade, which is usually imperceptible. The power of electromagnetic waves could be exploited for treatment on acupuncture points. The resulting stimulation affects the nervous and endocrine systems to provide dramatic biological results.

The important peculiarity of low intensity millimeter wave interaction with biological objects is that the result is strongly dependent upon frequency, i.e. it is observed a strong effect at resonating frequency and the effect is absent at another frequency [1]. There are many resonance frequencies in mm-band (e.g. it was observed more than 400 such frequencies in wave range from 5.5 to 7.5 mm). The different resonance frequencies act differently on biological objects.

Therapeutic Potential of Millimeter Wave Therapy

MRT is proving itself to be a remarkable therapeutic regimen. The reported success rate of MRT for various pathologies is astonishingly high. The experiments on the influence of mm waves on biological objects show that these waves are capable of affecting the basic vital functions of living cells. MRT improves metabolism of cells, restores microcirculation, optimizes exchange processes, normalizes peripheral innervations, have anti-spasm effect and normalize the function of an organ irrespective of an initial condition both for hypo and hyper function. MRT can be used as monotherapy [2] as well as in conjunction with other methods like surgery and drug treatment. MRT can be used in following conditions

- Peripheral nervous system: Neuralgia, neuritis, radiculitis, plexitis, plexalgia.
- Vegetative nervous systems: Sympalgia, ganglionitis, solaritis, neurocircular dystony, migranine, neurosis.
- Circulatory system: Hypertonic diseases, reflector stenocardia, varicous dilation of the upper extremity.
- Respiratory system: Chronic bronchitis, laryngitis, tracheitis, rhinitis, sinusitis, bronchial asthma.
- Digestive system: Gastric and duodenal ulcer, chronic gastritis and gastroduodenitis, uninfectious colitis, functional disorders of a stomach and intestines, chronic uncalculous cholecystitis, pancreatitis.
- Musculo-skeletal system: Various disorders by osteochondrosis of vertebral column and connected with it disorders.

Advantages of MRT are that, it is a contact free, infection free, AIDS safe, non-invasive method, having no side effects. It may be used in out patients' clinics, low in cost & there is no need to sterilize medical instruments. It has no adverse effects for the personnel.

Effect of Millimeter Wave Therapy

Sedative and Analgesic Effect

These are the most common effects of millimeter wave therapy cited by majority of physicians and patients. Usually after the first 2-3 sessions of millimeter wave therapy, 73-100% patients report alleviation of even total relief from the pain accompanying the disease, whether peptic ulcer [3,4,5,6] heart disease or prutic skin condition. This is followed by normalization of sleep improvement of general condition. Efficacy of millimeter wave therapy in treatment of males with psychogenic sexual dysfunction.

Some recent experiment confirmed that millimeter wave therapy is capable of interacting with neurons affecting the electrical characteristics of some neuro-peptides. A clinical study of 70 opoid drug abusers that millimeter wave therapy alone can significantly improve the condition of patients suffering from with-drawal symptom. Drug abuser reported the sensory feeling comparable with those from drug.

Anti-inflammatory Action

Early experiments with wounded rabbit, mice and other laboratory animals showed that the exposure of a wound surface to millimeter wave significantly enhance the rate of the recovery process by 1.5-2 times by decreasing duration oedema and exudative-inflammatory phase of wound healing. It has been reported that millimeter wave, when applied directly to infect wound surface cause a sharp decrease in microbial contamination of wound and favourable changes in sensitivity of micro-organism to some antibiotics. Former can be the result of (1) direct action of millimeter wave on micro-organism and/or (2) enhancement of host immune system. A decrease of R-plasmid-mediated resistance of *E. coli* to tetracycline has been observed *in vitro* and is potentially a very important feature of millimeter wave. It would be special value when applying millimeter wave for treatment of infected surface wounds (trauma, burn patient etc.). Millimeter wave therapy speeds up the recovery of patients suffering from various kinds of infected clean wounds and fractured bones. Millimeter wave therapy is also used for complicated bone fractures even with osteomylitis. In addition to treating the wound infections, millimeter wave therapy is reported to an effective method of preventing post surgical infections in cancer patients. The ability of millimeter wave to cause healing of skin without scarification was also demonstrated by dermatologists who treated skin wounds and lesion.

Immune System Stimulation

Laboratory investigation has confirmed that millimeter wave therapy produces non-specific enhancement of the human immune system. The changes include increased phagocyte activity of macrophages enhance proliferation and normalization of the ratio of CD+/CD8+T-lymphocytes, increased amounts of B-lymphocytes and normalized production of immunoglobulins. Millimeter wave therapy caused enhanced antimicrobial immunity resulting in faster clearance of affected organ/tissue from pathogens as in several other experimental and clinical studies. Normalization of the CD4+/CD8+ ratio of T-cells was observed in blood of patients with cardiac, diabetic, oncological and other pathologies. In response to millimeter wave therapy T-lymphocyte function and number both improved. Enhanced immune reaction mediated primarily through T-cell was found in mice exposed to millimeter wave. The above combination of features, confirmed, would make millimeter wave therapy a very powerful treatment modality.

The Millimeter Wave Therapy Method

Millimeter wave therapy consists of exposure of certain area of skin to low intensity millimetre wave. Any one of several site of application appears to be effective. Good results have been reported after irradiating various location over sternal area, large joints such as the shoulder or hip, some area of head such as occipital area or pineal gland projection area in the middle of forehead, biologically active zone and acupuncture points. In any event, with the exception of local treatment of skin disease or open surface of wounds the affected organ/tissue is usually remote from the site of application of millimeter wave.

The choice of frequency of millimeter wave is based on the two alternative principles. According to first one backed by N. Devyaktkov, M. Golant and O. Betskii three therapeutic wavelengths of 4.9, 5.7 and 7.1 mm (respective frequencies of 61.22, 53.57 and 42.25 GHz) were established as those that produce healing effect without harming the patient and were approved by Russian Ministry of Health. The second principle was introduced by S. Sitko. This approach is based on tuning the frequency output of millimeter wave generator usually within 53-67 GHz range, according to sensory response of a patient. A tunable millimeter wave generator or a device which produced a wide-band noise signal with an extremely low power output was used. The site of exposure of millimeter wave preferred by these researchers is an acupuncture point or points.

A typical therapeutic session lasts for 15 to 30 minutes, one exposure session per day, with a course comprised of 10 to 15 sessions depending on the nature and state of disease. Initially, existing industrial millimeter wave generators designed specifically medical purposes. These devices are capable of producing either a tunable fixed band output or sweeping signal within a set range of frequencies. Later, generators designed specifically for medical purposes appeared. They produced either a fixed frequency signal or broad band low power noise in the millimeter range. During millimeter wave therapy, the average power density incident to skin of patient is below 0.01 mW/cm². These features of medical millimeter wave generator are characteristic of newer devices as well, and prospective model are miniaturization and higher level of computer control of wavelength, duration of exposure and continuous or modulated signal output with the demand.

Conclusions

The field of microwave resonance therapy (MRT) has tremendous potential. The upper energy threshold of non-thermal bioeffects is about 10 mW/cm² not causing heating of bio medium of more than 0.1. Therefore, millimeter wave therapy did not produce thermal bioeffect. The 10 mW lower power generators are powered by 220 ± 22V 50 Hz AC and have output power density of 0.2-5.0 mW/cm much lower than biologically limited 10 mW/cm² during 8 h, as prescribed by Russian and Ukrainian National Standards. The output power density as well as the duration of the treatment, significantly influences the absorbed millimeter wave dose and corresponding millimeter wave therapy bioeffect, which can be biostimulative for the low-level therapeutically recommended dose of typically 20 minute daily millimeter wave therapy treatments (causing local temperature increases up to 38C, and maximally fast bioeffects). Millimeter wave therapy within the limits of frequency and power customarily used for medical purposes is safe. Evaluation of clinical and economical efficiency of millimeter wave therapy in the former USSR over almost a decade has included more than 50,000 patients with over 60 different pathologies from many institutions.

The following results were obtained:

1. Treatment efficiency was high (60-75%) depending on the kind of disease, state of evolved pathological process, and individual differences.
2. The healing process is shortened 1.5-2 times, with no significant side effect. Thus, millimeter wave therapy has a high clinical and economical efficiency.

References

1. V.A. Kichaev. EHF-therapy and its application in medicine practice.
2. M.A. Rojavin and M.C. Ziskin. Medical application of millimeter waves, Richard J. Fox Centre for Biomedical Physics, Temple University School of medicine, Philadelphia, USA.
3. Filippov I, Illarionov Y, Zalevsky V, Petij S, Ardelyan V, Mosiychuk L, Demeshkina L, Sergeychuk V, Lebedinsky Y. Modern methods of non-traditional therapy of peptic ulcer. *Likarska Sprava* 1996; **10-12:** 14-20.
4. Kutsenok VA. Microwave resonance therapy of gastric and duodenal ulcer. In: Fundamental and applied aspects of the use of millimeter electromagnetic radiation in medicine. Kiev, 1989: 192-3.
5. Vinogradov VG, Kisel LK, Mager NV. Results of treatment of gastric and duodenal ulcer with millimeter electromagnetic waves. *Vrachebnoe Delo* 1993; **1:** 85-7.
6. Starodub EM, Samogalska OE, Markiv IM, Luchanko PI. Effect of electromagnetic radiation of the extremely high frequency on the course of peptic ulcer associated with *Helicobacter pylori*. *Likarska Sprava* 1994; **1:** 85-7.

Topics in Electromagnetic Waves: Devices, Effects and Applications
Edited by J. Behari
Copyright © 2005, Anamaya Publishers, New Delhi, India

11. Role of Low Level Microwave Radiation on Cancer Development in Mice

R. Paulraj and J. Behari

School of Environmental Sciences, Jawaharlal Nehru University,
New Delhi-110 067, India

Abstract: In this study 7-8 week old male Swiss albino mice were selected. Experiments were conducted at two stages of carcinogenesis (i.e. promotional and progression effects). In one set single dose of 7,12- dimethylbenz(a)anthracene (DMBA) 100 μg/animal was applied topically on the skin of each animal. After two weeks of gap they were exposed to 9.9 GHz radiation 2 h/d, 3 days a week for a period of 16 weeks. Another set of animals were transplanted (ip.) with ascites 8×10^8 (Ehrlich-Lettre Ascites, strain E) carcinoma cells per mouse. These animals were exposed to above field for 14 days.

Mice exposed to above field after the topical application of single dose of DMBA did not develop any tumor. Swiss albino mice were transplanted intraperitoneally with ascites (8×10^8 cell/mouse) and exposed to above mentioned fields for 14 days did not show significant difference in the mortality among the control and exposed group. This shows ineffectiveness of microwave radiation after DMBA exposure.

Introduction

Cancer is the most dreadful disease that can strike any organism of multicellular grade of organization. Thus, humans, all higher forms of animals and plants can become potential victims of this nuptious disease. The neoplastic disease can occur in any tissue of the human body.

The process of carcinogenesis is a very complex one. It is generally considered as multifactorial, multigenic and multiphasic in nature. It is multifactorial in nature because it can be elicited by more than one causative agent. The process of carcinogenesis involves a sequence of several events. Operationally the whole process can be compartmentalized into initiation, promotion and progression. These events involve either genetic or epigenetic mechanisms. Initiation can be brought by chemical, physical and viral agents.

It is possible that the carcinogenic agents cause insult to the genetic make up of the target cell and this would subsequently cause changes in the proto-oncogenes as well as tumor-suppressor genes. These changes in the genes bring about changes in the growth factor or growth factor receptors. Consequently, the proliferative activities change would lead to neoplastic condition.

Several epidemiological studies over the past years have suggested an association between industrial and domestic exposure to low level environmental electromagnetic field and an increased risk of cancer (Wertheimer and Leeper,

1982; Wertheimer et al., 1995). In addition, the incidence of leukemia, brain tumors, eye melanoma, neuroblastoma etc. especially in children living in the vicinity of high voltage transmission lines have also been reported (Wilson et al., 1990; NRPB, 1992; Feychting and Ahlbom, 1993; Olsen et al., 1993; Verkasalo et al., 1993).

The ELF magnetic field cannot act as a cancer initiator as the interaction energy is too weak to damage the DNA to a large extent. However, it acts as a promoter or co-promoter. Earlier *in vitro* studies have reported that magnetic field induces the alteration in cellular functions, proliferation and activities of certain growth related enzyme such as ODC and PKC (Blackman et al., 1993; Luben et al., 1994). Stuchly et al. (1992) reported that 60 Hz magnetic field acts as co-promoter in a two stage skin carcinogenesis and suggest that the magnetic field does not alter the neoplastic number but accelerates tumor growth. In the long-term study by Chou et al. (1992), Sprague Dawley rats were exposed to pulsed 2450 MHz (SAR, ranging from 0.1 to 0.4 W/kg) for 21.5 hr/day, 7 days/week for 25 months. Their results also indicated that the irradiated animals had a higher incidence of tumor than sham irradiated control. Wu et al. (1994) studied the effect on Balb/C mice, which were irradiated with 2450 MHz RFR at $10mW/cm^2$ for 3 hr/day for 5 months. During this time dimethylhydrazine (DMH) was injected once in a week. Results showed no significant difference in colon cancer incidence between control and the treated one.

There are several reports documenting the incidence of cancer in workers exposed to MW, devices such as hand held mobile phones, hand held police radar guns, have been suggested to be associated with development of cancer such as brain and testicles. Epidemiological studies showed increased risk of brain cancer in those who engaged occupationally to such fields. WHO reports (WHO, 1993) state that low level RF fields are not mutagenic and may not be capable of initiating cancer in a manner similar to ionizing radiation and many chemical carcinogens. Though these fields do not initiate cancer, they are understood to affect carcinogenesis at the promotion level (McKinlay, 1993). Lai and Singh (1996) reported that microwave radiation cause both single and double strand breaks in DNA molecules. Somatic mutation in persons occupationally exposed to microwave has been reported (Garaj-Vrhovac et al., 1987). Skin tumors in mice provide a well-defined and accepted model to study multistage carcinogenesis and the potential effects of promoters on tumor growth McLean et al. (1991) studied the initiation and promotion of skin cancer in mice.

In the present investigation we have examined the effect of square wave modulated microwave radiation (9.9 GHz) on mice sensitized to tumor initiated by DMBA. This is to examine the sensitivity to the tumor promotion and the progression.

Methods

Skin Tumor Induction

Swiss albino mice (7-8 week old male) were obtained from the Jawaharlal Nehru University animal facility and were supplied with food and water *ad libitum*. These animals were divided into different groups. These mice were clipped off on the dorsal side (2 cm^2) and left for 3 days to check hair regrowth. Animals that showed hair regrowth were discarded.

The mice were kept in the cages (43 × 27 × 15cm) made of plexi glas and exposed in an Anechoic chamber. The cages were irradiated at the mentioned power level, which was of the type used previously (Ray and Behari 1990). Each cage contained 9 mice.

Exposure was given for 2 hr/day, 3 days a week for 16 weeks at 9.9 GHz radiation (square wave modulated) at power level 0.125 mW/cm^2. In these conditions we have adopted the specific absorption rate (SAR) value from the theoretical estimations of Durney et al. (1985) for the given exposure level. For a small mouse, where the E field can be assumed parallel to the body of the animal, SAR turns out to be 0.1 W/kg.

The sham exposure was identical to exposed group animals, but the system was not energized. Control mice (nine) were kept in an identical cage (43 × 27 × 15cm) which was also made up of plexi glas. The cage was placed inside the chamber for 2 h/day, 3 days a week for 16 weeks.

Experimental Design

The animals were divided into different control and experimental groups.

Group 1 ($n = 18$) mice acted as control (since earlier work in our laboratory did not produce any tumors in mice exposed topically to acetone we did not maintain this additional group here).

Group 2 ($n = 18$) mice were treated topically on the shaven area with a single dose of 7,12-dimethylbenz (a) anthracene (DMBA) (100 μg/100 μl acetone) and kept for 16 weeks.

Group 3 ($n = 18$) mice were treated as group 2 and after two weeks of gap they were exposed to square wave modulated 9.9 GHz microwave radiation with the power density of 0.125 mW/cm^2, 2 h/day, 3 days a week for a period of 16 weeks.

Group 4 ($n = 18$) were exposed to square wave modulated 9.9 GHz microwave radiation with the power density of 0.125 mW/cm^2, 2 h/day, 3 days a week for a period of 16 weeks.

Group 5 ($n = 18$) A single dose of DMBA (100 μg/100 μl acetone per animal) was applied topically on the dorsal side of the animal. After two weeks gap croton oil (1% oil in 100 ml acetone/animal) was applied topically twice a week for 16 weeks.

Tumor observation and body weight were conducted at weekly intervals in each group.

Effect of EMF on Progression

Seven to eight weeks old Swiss albino male/female mice were selected. These animals were injected intraperitoneally with 8×10^8 ascites carcinoma cells per mouse (Ehrlich-Lettre Ascites, strain E). They were divided into different control and experimental groups.

Control group (ascites bearing mice) ($n=8$) animals were sham irradiated to 9.9 GHz for 2 hr/day for a period of 14 days. Whereas experimental group (ascites bearing mice) ($n=8$) mice were exposed to 9.9 GHz for 2 hr/day for a period of 14 days.

After 14 days ascitic fluid was aspirated from the peritoneal cavity by a 5 ml syringe. Body weight was observed at weekly intervals. Cell count was performed under microscope.

Results

There was no spontaneous occurrence of skin-tumors in Jawaharlal Nehru University (of Swiss albino mice). When DMBA is topically applied on the skin, a single sub-carcinogenic dose also does not elicit skin tumors. However, when application of DMBA is followed by the long-term application of croton oil, the skin develops papillomas, at the end of the observation period. We have examined the effect of microwave irradiation at two levels. First in the promotional level and the second at the progression level.

Skin Carcinogenesis (Promotional Level Effect)

Treatment of mice with carcinogen (DMBA) and when independently irradiated with microwave frequency (9.9 GHz) did not result in loss of body weight. Mice treated topically with 100 µg DMBA (single dose) along with the radiation dose did not show any tumor development. This was confirmed visibly (Table 1).

Table 1. Effect of square wave modulated 9.9 GHz radiation on DMBA initiated mice

Group No.	Treatment	Number of animals		Average latent period (weeks)	Total tumor bearing animals
		Initial	Effective		
1	Control	18	16	00	00
2	DMBA	18	17	00	00
3	DMBA + 9.9 GHz	18	14	00	00
4	9.9 GHz	18	16	00	00
5	DMBA + Croton oil	18	15	14.25	10

Effect of EMF on Ascites (Progression Effect)

In another group, animals were transplanted intraperitoneally with ascites (8×10^8 cell/mouse) and exposed to above mentioned radiations for 14 days. In these groups animal could not survive for more than this period. There was no significant difference in the mortality among the control (ascites bearing mice) and exposed group (ascites bearing mice + irradiation). We did not observe any significant difference in the body weight (Fig. 1).

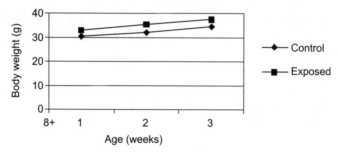

Fig. 1   Body weight of ascites bearing mice exposed to square wave modulated 9.9 GHz radiation.

The number of cells in mice transplanted with ascites only was 3.625×10^8 cells/ml, whereas in mice transplanted with ascites plus 9.9 GHz radiation group it was 3.875×10^8 cells/ml. Here also the difference was statistically insignificant ($p < 0.5$) (Fig. 2).

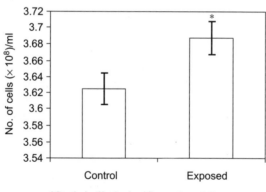

*Statistically insignificant ($p < 0.5$)

Fig. 2   Number of ascites cells in mice exposed to 9.9 GHz wave radiation.

Discussion

Since skin tumor is well established and one can easily measure the growth and number macroscopically and therefore, adopted skin cancer model for our study. Our results showed no tumor formation in the animals exposed for a period of 16 weeks in the above fields. In a similar experiment, Verschaeve and Maes

(1998) reported that mobile phone carrier frequency is not genotoxic. They do not induce genetic effects *in vitro* or *in vivo* and not teratogenic nor carcinogenic. This is because RF and microwave lack necessary photon energies to disrupt chemical bonds. The threshold for the disruption of chemical bonds by electromagnetic field occurs at photon energies around 7-10 electron volt, much above the energy of a radiofrequency/microwave photons.

Frie et al (1998) reported that C3H/HeJ mice were exposed for 78 weeks to 2450 MHz radiation and found no significant difference between exposed and sham-exposed groups with respect to the incidence of palpated mammary tumors, latency to tumor onset, and rate of tumor growth. In another study by Toler et al (1997) female cancer-prone mice (C3H/HeJ) were exposed to 435 MHz microwaves for 21 months (22 hr/day, 7 days/week). Under the conditions of the study, microwave exposure did not affect the incidence of mammary tumors, latency to tumor onset or survival when compared to sham exposure. Adey et al. (2000) reported that there was no tumor incidence on the rats exposed to frequency modulated (FM) microwave fields for a period 731-734 days.

Occupational exposure to RF microwave fields has been associated with increased brain tumor incidences and was significantly elevated to 10-fold among those used for 20 years or more (Thomas et al, 1987). Earlier reports suggest that magnetic field acts as a promoter or co-promoter in the two-stage skin carcinogenesis model (Stuchly et al., 1991, 1992).

In the present study, when mice were transplanted with ascites and exposed to above fields, there was a small increase in the number of cells in the exposed group as compared to control group. It may be that a further prolonged exposure to these fields along with a less number of cells implanted would enhance the growth of ascites.

The studies conducted so far (leukemia/lymphoma models) were used to examine the co-promotional effects of magnetic fields (Sasser et al., 1996). Stuchly et al. (1992) reported that the magnetic field does not alter the neoplastic number but accelerates tumor growth. These fields alter the activity of growth-related enzymes such as ornithine decarboxylase (ODC) and protein kinase C (PKC) as well as increase in cell division and proliferation *in vivo* (Adey et al., 2000; Paulraj and Behari, 2002, 2004). These parameters are indicative of a possible role in tumor promotion or co-promotion, though exact mechanism is yet to be elucidated. It is conceivable that these fields are responsible to enhance the growth and development of cells.

In our study the animals after being induced with cancer cells (ascites) could not survive for more than 14 days because of the virulent nature of the ascites. The subsequent exposure to the irradiation lasts for the same period. A small increase in the number of cells in the exposed group as compared to control group may be a pointer to a possibly additive effect. Form this present study it may be concluded that these field modulate carcinogenesis in mice. Since the exposure to RF/MW is identified as non-specific stress phenomena, it is possible that its control is through hormonal secretions and does not act directly onto the initiated cells. For this reason it is possible that an immediate effect is not visible.

References

Adey WR. Byus CV, Cain CD, Higgins RJ, Jones RA, Kean CJ, Kuster N, MacMurray NA, Stagg RB and Zimmerman GJ (2000). Spontaneous and Nitrosourea-Induced Primary Tumors of the Central Nervous System in Fischer 344 Rats Exposed to Frequency Modulated Microwave Fields, *Cancer Res* **60**: 1857–1863.

Durney CH, Massoudi H and Iskander MF (1985). Radiofrequency Radiation Dosimetry Handbook (fourth edition) Salt Lake City, UT, p 6.16.

Feychting M, and Ahlbom, A (1993). Magnetic fields and cancer in children residing near Swedish high-voltage power lines, *American Journal of Epidemiology* **138**: 467–481.

Frie JR, Jauchem JR, Dusch, JS, Merritt, JH, Berger RE and Stedham MA (1998). Chronic, Low -Level (1.0 W/kg) Exposure of Mice Prone to Mammary Cancer to 2450 MHz Microwaves, *Radiat Res.* **150**: 568–576.

Garaj-Vrhovac V, Horvat D, Brumen-Mahovic V, and Racic J (1987). Somatic mutations in persons occupationally exposed to microwave radiation, *Mutat. Res.* **181**: 321 (Abst).

Kavet R (1996) EMF and Current cancer concepts, *Bioelectromagnetics* **17**: 339–357.

Lai, H. and Singh, N.P. 1996 single and double strand breaks in rats brain cells after acute exposure to radio frequency electromagnetic radiation, *Int. J. Radiat. Biology* **69**: 513–521.

Luben RA, Morgan AP, Carlson A (1994). One gauss 60 Hz magnetic fields modulate protein kinase activity by a mechanism similar to that of tumor promoting phorbol esters. Bioelectromagnetics Society, 16th Annual Meeting. Proceedings, p. 74.

McLean JRN, Stuchly MA Mitchel REJ et a. (1991) Cancer promotion in a mouse- skin model by a 60-Hz magnetic field: II. Tumor development and immune response, *Bioelectromagnetics* **12**: 273–287.

McKinlay A (1993) Guidelines on limiting exposure to electromagnetic fields, *Radiol. Protect. Bull.* (UK) **148**: 19–24

NRPB (1992). Electromagnetic fields and the risk of cancer, Documents of the NRPB,vol.3 no. 1 (Chilton UK: National Radiation Protection Board).

Paulraj R and Behari J (2002). The effect of low level continuous 2.45 GHz wave on brain enzymes of developing rat brain, *Electromagnetic biology and Medicine* **21**: 221–231

Paulraj R and Behari J (2004) Radiofrequency Radiation effect on Protein Kinase C Activity in rats' brain, *Mutation Res* **545**: 127–130.

Ray S and Behari J (1990) Physiological changes in rats after exposure to low levels of microwaves, *Radiation. Res.* **125**: 199–201.

Sasser LB Morris JE and Miller DL (1996). Exposure to 60 Hz magnetic fields does not alter clinical progression of LGL leukemia in Fischer rats, *Carcinogenesis* **17**: 2681–2687

Stuchly MA, Lecuyer DW and McLean J (1991). Cancer promotion in mouse-skin model by 60-Hz magnetic fields 1, Experimental design and exposure system, *Bioelectromagnetics,* **12**: 261–271

Stuchly MA, McLean JRN, Burnett RGM, Lecuyer DW and Mitchel REJ (1992). Modification of tumor promotion in the mouse skin by exposure to an alternating field, *Cancer Lett.* **65**: 1–7.

Toler JC, Shelton WW, Frei MR, Merritt HR and. Stedham MA (1997) Long-Term, Low-Level Exposure of Mice Prone to Mammary Tumors to 435 MHz Radiofrequency Radiation, *Radiat Res.* **148:** 227–234.

Verschaeve L and Maes A (1998). Genetic, Carcinogenic and teratogenic effects of radiofrequency fields. *Mutat. Res.* **10:** 141–165.

Wertheimer N and Leeper E (1982) Adult cancer related to electrical wires near the home *Int. J. Epidemiol.* **11:** 345–355.

Wertheimer N Savitz DA Leeper E 1995 Childhood cancer in relations to indicators of Magnetic Fields from ground current sources, *Bioelectromagnetics* **16:** 86–96.

WHO (1993). Environmental health criteria 137, Electromagnetic Fields (300 Hz-300 GHz). Geneva World Health Organisation, pp. 15.

Wilson BW, Wright CW, Morris JE et al. (1990). Evidence for an effect of ELF electromagnetic fields on human pineal gland function. *J. Pineal Res.* **9:** 259–269.

Topics in Electromagnetic Waves: Devices, Effects and Applications
Edited by J. Behari

12. Reconstruction of Tomographic Images of Breast Phantoms with and without Attenuation Compensation

D. Kumar and Megha Singh

Biomedical Engineering Division, Indian Institute of Technology, Madras,
Chennai - 600 036, India

Abstract: Tissue phantoms of conical shape (diameter: 8.0 cm; height: 8.0 cm), similar to human breast, are simulated. The optical parameters of base and inclusions of this phantom are similar to that of goat kidney and adipose and spleen, respectively, mimicking tumor and fluid filled cyst. For tomography reconstruction at various heights of phantom the parallel beam configuration, with sixteen source regions with same number of photons and sixteen collecting regions, is considered. By Monte-Carlo simulation the normalized transmitted intensity projections during the interval of time window of 100-300 ps are considered and based on these the tomographic slices of various diameters are reconstructed. By this process the inclusions at slice diameters 2.0 and 3.0 cm are only visualized. To retrieve the data on inclusions at various levels, two different procedures, either by image processing procedures or by attenuation compensation, are implemented. The size of the inclusions in slices of diameters 2.0-6.0 cm obtained by attenuation compensation show lesser deviation (0.0 to 2.5%) compared to that obtained by image processing procedures (10-75%), respectively.

Introduction

Optical characterization of biological tissues is of growing importance as this procedure is non-invasive and non-ionizing. After its application to demonstrate the characteristic absorption by the oxygenated and the deoxygenated forms of hemoglobin at optical wavelengths [1], various procedures are developed and applied in detection of physiological, structural and metabolic changes in tissues such as early detection of human breast tumor and abnormal cerebral oxygenation or perfusion to prevent permanent brain injury [2-4]. In spite of reasonably well penetration of light in thick biological tissues at red and near infra-red (NIR) wavelengths, the reconstructed images are noisy due to high scattering. To minimize this effect, the collection of collimated photons or minimally scattered photons is required. For this purpose, the techniques based on time-resolution [5], polarization gating [6] and frequency-domain [7] are applied in tumor detection in tissues. To minimize the effects of light scattering and diffusion in biological tissues the tomographic imaging from multiple projections is implemented [8]. Despite this the images obtained at large slice diameters are associated with noise due to multiple scattering of transmitted photons [9].

For better description of the process of photon-tissue interaction both for monitoring and imaging, theoretical models are necessary to describe photon migration through highly scattering media. Radiative transfer equations (RTE) describe the light transport or the randomized interaction of photons with the tissues [10]. For more complex geometry or inhomogeneous medium the solutions of RTE are derived by the finite-element method [11] or Monte-Carlo simulation (MCS) [12], which provides light fluence distribution inside a tissue medium. The theoretical studies carried out by the MCS reveal the feasibility of time-gated transillumination imaging system [10], resolution of imaging through random media [13] and accurate recovery of images [14].

During the recent years, optical tomography has emerged as a technique to monitor various tissue changes. This is primarily attributed to generation and distribution of light and data acquisition for optical tomography. The image reconstruction from light projections, however, is a difficult task because of strong scattering and diffusion of light in the biological tissues. For this purpose, time-resolved profile [15] and modified tomographic reconstruction techniques [16], compensated transillumination [17], model based image reconstruction schemes [18] and data correction algorithm to minimize the effect of background heterogeneity [19] are developed. Despite these developments, in the slices of larger diameter breast like phantoms, the image contrast and resolution are reduced, whereas, the inclusions in the slices of small diameter are clearly seen [9].

For better visualization of the inclusion in the tissues in slices of larger diameter, further developments are required. The improvement in these images could either be achieved by processing of the slice image by a set of image processing procedures or to compensate the loss of photons in large diameter slices to bring the image quality similar to that of smaller slices in which the inclusions are clearly observed. Both procedures are to be applied onto the simulated images as obtained by MC simulation of transmitted photons within time window (100-300ps). This forms the objective of the present work to develop these procedures for better differentiation of various abnormalities in healthy soft tissues.

Materials and Methods

Monte-Carlo Simulation

For tomographic reconstruction the transmission profile through the phantom was simulated by Monte-Carlo procedure [20]. Fig. 1 shows the schematic of the simulated propagation path of a photon, after incident normally on the tissue surface at (0, 0, 0). The incident photon beam of diameter 0.1 cm and each photon with unit intensity (WT_1) are considered.

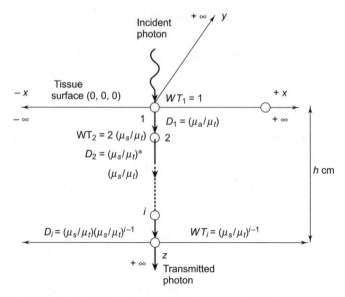

Fig. 1   Schematic of Monte-Carlo simulation for transmission of photon. o : scattering
centre, WT_i : weight at ith scattering position, D_i : absorbed dose during ith
scattering and h : sample thickness.

Since it is a normal incidence, the photon moves along the direction of incidence
without deflection and reaches the new position 2 from position 1. Due to tissue-
photon interaction the intensity is attenuated by the absorption process, which
is given by

$$\Delta Q = (WT_1)(\mu_a/\mu_t)$$

where μ_a is the absorption coefficient (cm^{-1}); μ_t the total attenuation coefficient
$(cm^{-1}) = (\mu_s + \mu_a)$ and μ_s the scattering coefficient (cm^{-1}).

The corresponding path length l in the tissue medium is given by

$$l = -(\ln R)/\mu_t$$

where R is a random number between 0 and 1.

Thereafter the new photon intensity WT_2 is calculated by

$$WT_2 \rightarrow WT_1(\mu_s/\mu_t)$$

After position 2, the photon is scattered and deflected.

The deflection angle θ is calculated by

$$m = \frac{1}{2g}\left[1 + g^2 - \left\{\frac{(1-g^2)}{(1-g+2g\varsigma)}\right\}^2\right] \text{ for } g \neq 0$$

where $g = \langle \cos \theta \rangle$ and ς is a random number between 0 and 1.

For $g = 0$

$$m = (2\varsigma - 1)$$

The selection of azimuthal angle ψ is given by

$$\psi = 2\pi\gamma$$

where γ is a random number between 0 and 1.

The propagation of photon in the tissue medium continues till its intensity reduces either to 0.1% of the incident intensity or the same photon is transmitted through the tissue surface. Thereafter, the next photon enters into the tissue and repeats the same process described earlier. By this process tracking of all the photons is carried out. The transmitted fractions of photons corresponding to various radial positions are represented as normalized transmitted intensity, given as percentage transmitted intensity of photons over the surface with respect to the incident (10^6) photons. The selected time of flight is decided in the simulation as the allowed pathlength between the source and the detector. Every transmitted photon should fly until it reaches the distance set by $c_n t_f$, where c_n is the speed of light in the medium, and t_f is the allowed time of flight.

Simulation Geometry

The model consisting of the sources, conical shaped sample and collecting regions is shown in Fig. 2. There are 16 sources of photons each of diameter 0.26 cm placed equidistant with 0.26 cm separation between two successive sources covering the full height (8.0 cm) of the conical sample of base diameter 8.0 cm. The transmitted photons after passing through respective locations are collected by same number of collecting regions (area of each region 0.26 cm × 0.26 cm).

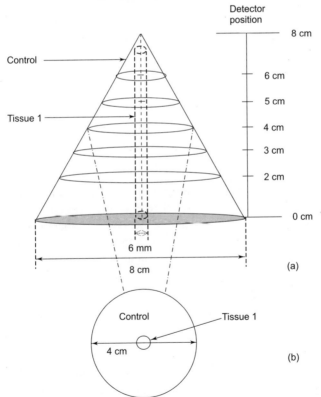

Fig. 2   Schematic of geometric description of the cone and the presence of cylindrical inclusion inside the cone (a) and the cross-section of the slice at a height of 4 cm from the base (b).

Thus, forming a parallel-beam configuration of tomographic process. The tissue phantom considered for this purpose is a cone with base diameter 8.0 cm (Fig. 3). The space between the cone and the source and the detector lines is filled with nonscattering and non-absorbing medium with the same refractive index as that of the cone material. The tomographic images are reconstructed at heights 2, 3, 4, 5 and 6 cm, which mean that the path length for all the photons is more than that of slice diameter. The attenuation due to these additional paths is not considered for tomographic reconstruction. The acceptance angle of each region is 0.1 rad. The time of flight of photons varies from 100 to 300 at slice diameter 2 cm to 6 cm, respectively.

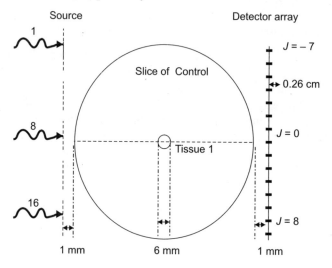

Fig. 3   Schematic of photon source and collecting regions with the cross-section of control tissue and the presence of either tissue 1 or tissue 2.

To obtain each projection at a height 6.0 cm, a two dimensional grid of size 1×16 was considered for storing the transmitted photons from the surface. The photons collected within a region of 0.26 cm × 0.26 cm around the grid point were considered to be associated at that location $A_{i,j}$ where i is the slice level and varies from 1 to 5, and j refers to detector location in horizontal position and varies from –7 to 8. Total number of photons incident on the tissue surface was considered as A_0. The normalized transmitted intensity (NTI) in terms of percentage is given by

$$\text{NTI (\%)} = \frac{A_{i,j}}{A_0} \times 100$$

A similar grid is used to store distribution of photons at slice diameters 5.0, 4.0, 3.0 and 2.0 cm and their projection profiles were obtained. Since cone and placement of the cylindrical inclusions at the center are symmetrical in nature, the one projection data obtained by simulation was utilized to get other projections for each angular increment of 3.6°.

Simulation of Tissue Phantoms

For simulations purpose the optical parameters of three tissues kidney, adipose and spleen, were used. The absorption μ_a and scattering μ_s coefficients and anisotropy parameter g of these, by matching of the surface reflectance profile with that as obtained by Monte-Carlo simulation, were determined. These parameters ($\mu_s/\mu_a/g$) of kidney, adipose and spleen tissues are 76.0 cm^{-1}/ 0.8 cm^{-1}/0.994, 419.9 cm^{-1}/1.5 cm^{-1}/0.994 and 109.8 cm^{-1}/4.0cm^{-1}/0.995, respectively. Further details of this procedure are given elsewhere [21].

Based on known optical parameters, the tissue models with compositional variation were developed. The cylindrical shaped spleen T_1 or adipose T_2 tissue of diameter 0.6 cm and length 7 cm was introduced vertically at the center of the base of conical phantom of control tissue (kidney). Refractive indices n of control C, T_1 and T_2 are 1.40, 1.40 and 1.455, respectively. Thus, providing the matched and mismatched boundary conditions between C and T_1, and C and T_2, respectively, which are considered for modeling purpose [22].

During irradiation of the C or C with inclusion T_1 or T_2, by collimated beams of photons, a part of each beam is minimally scattered within the medium. Another component of this emerges as transmitted fraction at various locations on the phantom's surface, which were collected. From this data the normalized transmitted intensity (NTI) or normalized attenuated intensity (NAI) are obtained. By similar procedure these components at various locations of the phantoms were determined.

Multi-slice Tomographic Imaging

This involves the following steps:

Interpolation: The data set of the NTI was at discrete locations on the surface. From this data by a linear interpolation process a regular pattern, a grid of 1 × 64 was obtained. Each point was characterized by its (x_i, y_i) coordinates and its NTI. Based on 100 projections the simulated data grid of size 100 × 64 was obtained.

Tomography reconstruction: The convolution back-projection algorithm applicable for parallel beam configuration was used to obtain the tomographic images at various heights of the phantom. The image reconstruction by this procedure is based on filtered-back-projection technique [23]. A brief outline of this procedure is given as follows:

Step 1: To obtain filtered back-projections, the measured projections $P_\theta(d)$ are convolved

$$Q_\theta(d) = P_\theta(d) \cdot h(d) \tag{1}$$

where $P_\theta(d) = \int_{ray} f(x,y)ds; f(x,y)$ is the two dimensional distribution function of the cross section of the slice; $h(d)$ is a filter function, $d = x \cos\theta + y \sin\theta$, and θ the angle between x-axis and a line perpendicular to the 'ray' running through (x, y).

Step 2: The original function is reconstructed by

$$f(x,y) = \frac{\pi}{M} \sum_{i=0}^{M} Q_\theta (x \cos\theta + y \sin\theta) \qquad (2)$$

where M is the total number of angles θ for which the projections $P_\theta(d)$ are known.

The projection data are convolved with a filter and the filtered projections are given by equation (1). By applying equation (2), the tomographic function is reconstructed. This function is in the form of a two dimensional array of numbers. This array is converted into a tomographic image by assigning gray levels from 0 to 255 and to present this data in percentage, the gray levels from 0 to 100 are assigned. Finally, the reconstructed image of size 64 × 64 pixels was displayed. To reduce the spike noise this image was further processed by median filtering. By the same procedure the tomograms of various diameters of the phantoms were obtained. These tomograms were scanned along the horizontal diameter and their attenuated intensity plots were made. To obtain the size of the inclusion placed inside the control, the tomographic images were further processed [24].

Sequence of Image Processing Steps

The details of the image processing procedures (Fig. 4) are given as follows:

Subtracted image: This is obtained by subtracting the control image from the image of the control with inclusion (spleen). By this the region of the influence of the embedded tissue is obtained. To obtain the details this needs to be further processed.

Contrast adjustment: This operation transforms an input image $A(x, y)$ into an output image $B(x, y)$ and is consistent for the whole image:

$$B(x, y) = f[A(x, y)]$$

In digitized images, each starting pixel is converted to another pixel at the same location which derives its gray level from the original pixel using law 'f'. It has been implemented in the input image such that this reduces the contrast in brighter

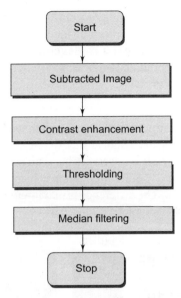

Fig. 4   Sequence of image processing steps.

areas and increases the contrast in darker areas of the image. With this process, the image resolution is improved.

Threshold: In this operation, certain level in histogram is specified as threshold. The pixels lighter than the threshold are converted to white and the pixels darker are converted to black. Thus, the gray image is transformed into a binary image leading to clearly distinguishable regions.

Median filtering: This procedure reduces noise in an image by blending the brightness of pixels within a selection. This involves a non-linear filter, and does not introduce any new data point. Typically these operations are concerned with nearest neighbors only, so that the new value of an element depends on nine previous values, i.e. the element itself and its eight nearest neighbors. For this the input data value was replaced by the median value, that is

$$v(m, n) = \text{median } \{y(m{-}k, n{-}l), (k{-}l) \in W \}$$

where W is a suitably chosen window of size 3×3. By moving the window, point by point, median was found. The center data point was replaced by the median value. Thus, the whole data matrix was filtered. By this filtering procedure the data values were free from noise and suitable for display.

Attenuation Compensation

The transmitted photons with a time window of 100 ps for a particular number of incident photons at slice diameter 2.0 cm of the conical phantom are considered as reference. With reference to this the required increases in input beam for slices of control tissue at various heights are calculated. By this procedure the same transmitted intensity at all the detectors for slices of varying diameters of the control is obtained. By the same procedure the incident photons at various diameters of control tissue with spleen as inclusion are calculated. Since the incident photons vary at different slice levels of the conical phantom, the computation time with Pentium IV processor varied from few hours to 3 days.

Results

Figure 5 shows the reconstructed cross-sectional images of control (a), control with spleen (b) tissue along the height at slice diameters 2 cm (i), 3 cm (ii), 4 cm (iii), 5 cm (iv) and 6 cm (v). It is observed that with increase in diameter of the slice for both control and $C + S$ the attenuation is increased. The attenuation is more in $C + S$ than control. The difference is maximum for the slice of diameter 2 cm and minimum at slice diameter 6 cm.

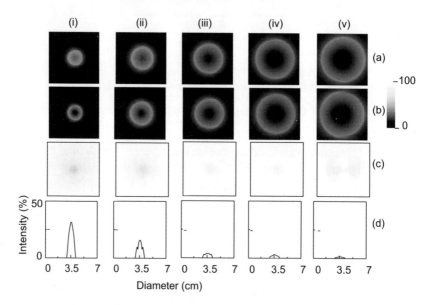

Fig. 5 Reconstructed cross-sectional images of control (a), control with spleen (b) tissue placed inside the conical geometry at the height of 6 cm (i), 5 cm (ii), 4 cm (iii), 3 cm (iv) and 2 cm (v) from the base, the subtracted images (c) and their associated scans (d).

The size of the inclusions obtained from above figures shows large deviations. To reduce the effect of multiple scattering, these images were processed by a set of image processing procedures. Figure 6 shows an example of the processed images of the tomographic slice at height 4.0 cm. By this procedure the sizes of the inclusions in slices of various diameters are determined.

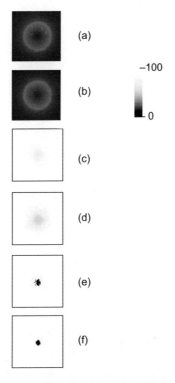

Fig. 6 Processing of images by various procedures of tissue phantom containing spleen as inclusion at the slice diameter of 4.0 cm. Unprocessed image of $C+S$ (a) and control (b), subtracted image (c), contrast enhanced image (d), thresholded image (e) and median filtered image (f).

Figure 7 shows the increase of incident photons with increase in slice diameter of the phantom. Since with the increase in slice diameter the attenuation increases, this should be compensated with the increase in number of photons at various location of diameter larger than 2.0 cm to obtain the constant transmitted ($\times 10$) photons at all the slices of the varying diameter. This shows that there is an exponential increase in incident photons with the increase in slice diameter.

Figure 8 shows the reconstructed cross-sectional images of control (a) and control with spleen (b) tissues at slice diameter 2 cm (i), 3 cm (ii), 4 cm (iii), 5 cm (iv) and 6 cm (v). Using the technique of attenuation compensation, it is observed that there is no variation in intensity in all the slices of the conical biological tissue without cylindrical inclusion. But with inclusion, the decrease in attenuation of intensity with increase of slice diameter is observed. The difference is maximum for the slice at height of 6 cm and minimum at 2 cm from the base.

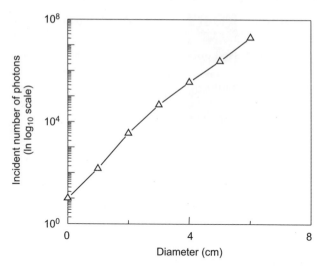

Fig. 7 Input photons required to obtain constant (×10) transmitted photons irrespective of slices of various thicknesses.

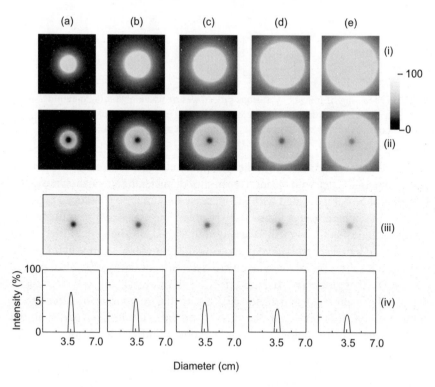

Fig. 8 Reconstructed cross-sectional images of control (a), control with spleen (b) tissue placed inside the conical geometry at the height of 6 cm (i), 5 cm (ii), 4 cm (iii), 3 cm (iv) and 2 cm (v) from the base, the subtracted images (c) and their associated scans (d) using the technique of attenuation compensation.

By the same procedure the tomographic images of control with adipose tissue are obtained. By attenuation compensation and image processing procedures these images are processed and the sizes of inclusions are determined. A comparison of the sizes of inclusions before and after image processing and with attenuation compensation is given in Table 1. As determined by using the technique of attenuation compensation the processed sizes are close to the actual sizes of the inclusion at various slice diameters.

Table 1. Determined size of spleen S or adipose A tissue located inside the control tissue at various diameters of the slices before and after processing the image without attenuation compensation, and with attenuation compensation (actual size of inclusion = 0.6 cm)

Slice diameter (cm)	Size (mm)				
	Before processing		After processing		Attenuation compensation
	$C + S$	$C + A$	$C + S$	$C + A$	$C + S$
2	6.6	7.6	6.6	7.6	6.0
3	6.6	7.6	6.6	7.6	6.0
4	8.7	10.9	6.0	7.0	6.0
5	8.7	10.9	7.0	7.0	6.15
6	10.9	10.9	8.7	9.2	6.15

Discussion

Biological tissues are highly scattering in nature because of their structural variation. The photons propagating inside the thicker tissues like breast are multi-scattered, which results in a noisy image. The time windowing is useful to minimize the effect of multiple scattering inside the tissue [25]. The time window of 100 ps to 300 ps, depending upon the slice diameter, enables to collect the ballistic, quasi ballistic and minimally scattered components of transmitted photons. Here, the time of collection of photons is too short, and the photon travel is considered in a straight line to apply convolution back-projection algorithm. This algorithm back-project the absorption values along the straight line between the source and detector [23]. Since light travels a longer distance at the center, the photons undergo finite scattering. The attenuation varies with the change in the light path. This is the minimum near the periphery and maximum at the center.

The higher attenuation due to the presence of spleen tissue is attributed to its higher absorption coefficient. With the inclusion of adipose it is expected that the transmission at the mid-position of the slice should be higher but due to the refractive index mismatch between C and A the trapping of photons causes higher attenuation [22]. The parameter full-width at half maximum (FWHM) in the subtracted images is used to provide comparative information on the size of the inclusion [26] but highly significant increase in this parameter is observed with the increase of slice diameter. It is attributed to the high scattering within the tissue medium. The applications of several image processing procedures improve the quality of the image for identification of the size of the inclusion [3, 5, 9, 25]. But the calculated size is still larger even after the processing. This could be attributed to the multiple scattering of less number of photons, after passing through the inclusion, in the large thickness of base material.

The high scattering in the tissue medium leads to intense lengthening of the path, which finally leads to increased probability of absorption in the object [27]. The optical attenuation, which depends on the scattering and absorption, increases with the increase of slice diameter and is given by the product of attenuation coefficient multiplied by zigzag pathlength [28]. This decrease in transmitted intensity is exponential as given by Beer-Lambert law. The available number of transmitted photons is very much less in the slices of higher diameters, as the photon traverses longer path length for their detection. By increasing the number of photons as input, the quantity of loss of photons during their travel is adjusted. For attenuation compensation technique the number of transmitted photons at all heights collected over a short time window (100-300 ps) over all the detectors is constant. The decrease in attenuation with increase in slice diameter is attributed to weak coupling of incident photons as the tissue of inclusion is moving deeper from the point of the incident photons. The proposed technique of attenuation compensation is found appropriate in many ways. First, because of keeping the constant number of transmitted photons, it provides uniform intensity distribution in all the slices of varying diameter while construction of tomograms. Second, the deviation from the constant transmission output is clearly noted when there is an inclusion in the control. Thus, during tomogram reconstruction the inclusion is distinctly observed. The contrast and resolution of the image is well improved. The images of better contrast could still be produced by reducing the acceptance angle and size of the collecting area.

Conclusion

The presence of inclusions in the tissues could be analyzed by simulation of tomography procedure. In contrast to image processing procedures, the better quality images of tissues and their inclusions are obtained by attenuation compensation technique. The application of this technique may further strengthen the existing knowledge of optical tomography in detection of abnormalities in human breast tissues.

References

1. Jöbsis, F. F. "Noninvasive, infrared monitoring of cerebral and myocardial oxygen sufficiency and circulatory parameters," *Science,* **198,** 1977, 1264–1267.

2. Okada E., Firbank, M., Schweiger, M., Arridge, S. R., Cope, M., and Delpy, D. T. "Theoretical and experimental investigation of near-infrared light propagation in a model of the adult head," *Appl. Opt.*, **36**, 1997, 21 – 31.

3. Ntziachristos V., Hielscher, A. H., Yodh, A. G. and Chance, B. "Diffuse optical tomography of highly heterogeneous media," *IEEE Trans Med. Imag.*, **20**, 2001, 470–478.

4. Svanberg S., "Some applications of ultrashort laser pulses in biology and medicine," *Meas., Sci., Technol.*, **12**, 2001, 1777-1783.

5. De Haller E. B. "Time-resolved transillumination and optical tomography," *J. Biomed. Opt.*, **1**, 1996, 7–17.

6. Chandran, G. G., Vasu, R. M. and Asokan, S. "Tomographic imaging of phase objects in turbid media through quantitative estimate of phase of ballistic light," *Opt. Comm.*, **191**, 2001, 9–14.

7. O Leary M. A., Boas, D A., Chance, B. and Yodh, A. G. "Experimental images of heterogeneous turbid media by frequency-domain diffusing-photon tomography," *Opt. Lett.*, **20**, 1995, 426–428.

8. Gao, F., Poulet, P. and Yamada, Y. "Simultaneous mapping of absorption and scattering coefficients from a three-dimensional model of time-resolved optical tomography," *Appl. Opt.*, **39**, 2000, 5898–5910.

9. Chacko, S. and Singh, M. "Three-dimensional reconstruction of transillumination tomographic images of human breast phantoms by red and infrared lasers," *IEEE Trans. Biomed. Eng.*, **47**, 2000, 131–135.

10. Arridge, S. R. "Photon measurement density functions. Part I. Analytic forms," Appl. *Opt.*, **34**, 1995, 7395–7409.

11. Arridge, S. R. and Schweiger, M. "Photon measurement density functions. Part 2: Finite-element method calculations," *Appl. Opt.*, **34**, 1995, 8026–8037.

12. Wilson, B.C. and Jacques, S.L. Optical reflectance and transmittance of tissues: Principles and applications. *IEEE J. Quant. Electr.* **26**, 1990, 2186–2199.

13. Chen, Ye. "Characterization of the image resolution for the first-arriving-light method," *Appl. Opt.*, **33**, 1994, 2544–2552.

14. Pogue B. W., S. Geimer, T. O. McBride, S. Jiang, Ulf L. Österberg and K. D. Paulsen, "Three-dimensional simulation of near-infrared diffusion in tissue: boundary condition and geometry analysis for finite-element image reconstruction," *Appl. Opt.*, **40**, 2001, 588–600.

15. Jarry G., Lefebvre, J. P., Debray, S. and Perez, J. "Laser tomography of heterogeneous scattering media using spatial and temporal resolution". Med. Biol. *Eng. Comput.*, **31**, 1993, 157–164.

16. Colak, S. B., Papaioannou, D.G., 't Hooft, G. W., van der Mark, M. B., Schomberg, H. J. C., Paasschens, J., Melissen, J. B. M. and van Asten, N. A. A. J. "Tomographic image reconstruction from optical projections in light-diffusing media," *Appl. Opt.*, **36**, 1997, 180–213.

17. Wu, X. and Faris, G. W.. "Compensated transillumination". *Appl. Opt.* **38**, 1999, 4262–4265.

18. Paithangar, D. Y., Chen, A. U., Pogue, B. W., Patterson, M. S. and Sevick-Muraca, E. M. "Imaging of fluorescent yield and life time from multiply scattered light reemitted from random media," *Appl. Opt.* **36**, 1997, 2260–2272.

19. Hielscher, A. H., Yodh, A. G., and Chance, B. "Diffuse optical tomography of highly heterogeneous media," *IEEE Trans. Med. Imag.* **20,** 2001, 470–478.

20. Kumar, D. and Singh, M. "Characterization and imaging of compositional variation in tissues," *IEEE Trans. Biomed. Eng.*, **50,** 2003, 1012–1019.

21. Kumar, D. and Singh, M. "Non-invasive imaging of optical parameters of biological tissues," *Med. Biol. Eng. Comput.*, **41,** 2003, 310–316.

22. Tualle J. M., Tinet E., Prat, J. and Avrillier, S., "Light propagation near turbid-turbid planar interfaces," *Optics Communications,* **183,** 2000, 337–346.

23. Kak, A. C. and Slaney, M. "Principles of Computed Tomography Imaging," IEEE Press, New York, 1987, 1345–1372.

24. Jain A. K., "Fundamentals of Digital Image Processing", Prentice - Hall, Engelwood Cliffs, N. J.,1989.

25. Andersson-Engels, S., Berg, R. and Svanberg, S. "Time-resolved transillumination for medical diagnostics," *Opt. Lett.*, **15,** 1990, 1179–1181.

26. Jiang H., Xu, Y., Iftimia, N., Eggert, J., Klove, K., Baron, I. and Fajardo, L. "Three-dimensional optical tomographic imaging of breast in a human subject", *IEEE Trans. Med. Imag.*, **20,** 2001, 1334–1340.

27. Key, H., Davies, E. R., Jackson, P. C. and Wells, P.N.T. "Monte Carlo modeling of light propagation in breast tissue," *Phys. Med. Biol.* **36,** 1991, 591–602.

28. Tsuchiya Y., Photon path distribution and optical response of turbid media: theoretical analysis based on microscopic Beer-Lambert's law. *Phys. Med. Biol.* **46,** 2001, 2067–2084.

Topics in Electromagnetic Waves: Devices, Effects and Applications
Edited by J. Behari
Copyright © 2005, Anamaya Publishers, New Delhi, India

13. Microwave as a Probe to Detect the Purity of Desi Ghee (Saturated Fats)

Shilpi Agarwal and Deepak Bhatnagar
Microwave Lab, Department of Physics, University of Rajasthan,
Jaipur-302004, India

Abstract: The dielectric parameters of a system containing desi and vanaspati ghee (saturated fats) in different proportions by volume are measured at two different frequencies and scales to check the amount of purity of a given sample is prepared. It is found that the proposed technique to check the purity of samples at microwave frequencies is better than the traditionally applied optical technique because in the proposed technique, several parameters may be determined at a time. Hence, better conclusions may be made.

Introduction

The dielectric properties of agri-food materials and their constituents describe their molecular interaction with electromagnetic energy and depend on the frequency of electromagnetic field as well as on the bulk and microscopic properties of the materials and their composition. It is, therefore, important to know the dielectric properties of materials for the development of microwave process and control. During survey in market it was noticed that desi ghee available in open market contains impurities in the form of vanaspati ghee, refined oils and butter oil etc. The percentage of impurities varies significantly. AGMARK, Jaipur and ISI offices have made their own specifications for rejecting the samples containing impurities. For checking impurities, traditionally chemical testing of samples and Abbe's Refractometer techniques are applied. In the present communication, microwave technique has been used and a scale to find the amount of impurities in agri-food materials is prepared which provides the information that the available product is fit for consumption or should be rejected.

Preparation of the Sample and Experimental Details

Samples of desi ghee were collected from the market and chemical testing of these samples was carried under the guidance of AGMARK to find the acceptable sample for consumption i.e. purest sample among the available samples. Vanaspati ghee as impurity was added in different proportions by volume. The experimental set up to determine relative dielectric constant ε' and dielectric loss ε'' of ghee samples at microwave frequency is based on Von-Hipple method for lossy dielectric liquids [1]. By applying experimental set-up and a method proposed by Yadav and Gandhi [2] for estimation and measurement of dielectric properties of powders at MW frequencies, the dielectric constant and dielectric loss of pure Ghee containing impurities in different percentages by volume are determined by applying following relations:

$$\varepsilon' = \left(\frac{\lambda_0}{\lambda_c}\right)^2 + \left(\frac{\lambda_0}{\lambda_d}\right)^2 \cdot \left[1 - \left(\frac{\alpha_d}{\beta_d}\right)\right]^2$$

$$\varepsilon'' = 2 \cdot \left(\frac{\lambda_0}{\lambda_d}\right)^2 \left(\frac{\alpha_d}{\beta_d}\right)$$

where λ_0, λ_c and λ_d are the free space wavelength, cut-off wavelength and wavelength in the dielectric sample, respectively; α_d and β_d are attenuation constant of the material measured in nepers per meter and phase shift per unit length of the sample measured in radians per meter, respectively, and are calculated by the relations

$$\alpha_d = \frac{2.302}{2L} \cdot \log\left[\frac{\sqrt{x_1}}{2\sqrt{x_2} - \sqrt{x_1}}\right]$$

$$\beta_d = \frac{2\pi}{\lambda_d}$$

x_1 and x_2 are output power readings without and with sample of thickness L in the wave-guide. Above equations are employed to obtain the values of ε' and ε'' of the ghee samples. The main quantities to be measured experimentally are α_d and λ_d for the samples for which ε' and ε'' values are to be obtained. Various ghee samples containing impurities in different proportions were placed in the wave-guide. Microwave power is obtained from microwave source and is allowed to propagate through sample, which forms standing waves in guide section after being reflected from the short-circuiting wave-guide. This standing wave pattern is used to find λ_d and α_d.

The precision of measurement for the wavelength with the available X-band microwave test bench is ± 0.001 cm. Corresponding to this accuracy value, the error in the measurement of ε' is estimated. For simplification, involved errors due to non-zero impedance of the short circuit plunger are ignored. The errors of measurement are calculated by using the conventional method of error analysis [3] which states that if a quantity Q depends on several observable quantities x, y,....., such that Q is a known function of variables, $Q = f(x, y,...)$, then an error q of the quantity Q may be obtained from

$$q = \sqrt{q_x^2\left(\frac{\partial f}{\partial x}\right)^2 + q_y^2\left(\frac{\partial f}{\partial x}\right)^2 + \cdots}$$

where q_x, q_y, $\cdots$ are the errors in the measured values of x, y, ... This relation is valid even if the precision of respective measurements differs [3]. With this method, the accuracy of the measurement for ε' is found to be $\pm 1\%$ while for ε'' it is $\pm 5\%$.

The dielectric constant $\varepsilon_\infty = n_D^2$ at optically high frequency is determined by squaring the refractive index n_D for the samples, obtained experimentally by using Abbe's Refractometer. The dielectric constant at static frequency ε_s is measured with the help of a dipole meter based on the principle of hetrodyne beat.

Results and Discussion

The various values of dielectric constant at microwave frequency, optical frequency and static frequency are shown in Table 1.

Table 1. Experimental values of dielectric constant, dielectric loss, conductivity and relaxation time

Name of sample	n_D^2	ε_s	ε_∞	ε'	ε''	σ	$\tau \times 10^{13}$ s
Pure desi ghee	1.455	3.05	2.11	2.463	0.058	0.031	3.86
Desi ghee + 20% vanaspati	1.456	3.10	2.12	2.473	0.090	0.052	6.53
Desi ghee + 40% vanaspati	1.457	3.16	2.12	2.491	0.097	0.082	7.32
Desi ghee + 60% vanaspati	1.457	3.20	2.12	2.502	0.180	0.097	7.45
Desi ghee + 80% vanaspati	1.459	3.23	2.12	2.514	0.210	0.012	8.22
Pure vanaspati	1.460	3.26	2.13	2.532	0].250	0.014	8.42

ε_s = dielectric constant at static frequency, ε_∞ = dielectric constant at optical frequency, ε' = dielectric constant at microwave frequency, ε'' = dielectric loss at microwave frequency, σ = conductivity, τ = relaxation time, n_D^2 = refractive index.

From the present investigations carried out under the instructions of AGMARK, Jaipur, the sample having refractive index between 1.4545 to 1.4555 may be accepted for consumption otherwise it should be rejected. Corresponding to this range of refractive index, dielectric constant ε_∞ at optically high frequency ranges between 2.115 and 2.118. From the measured results, the value of refractive index for purest desi ghee is 1.455 and dielectric constant ε_∞ at optically high frequency is 2.117. Therefore, this sample is acceptable for consumption as per AGMARK specifications.

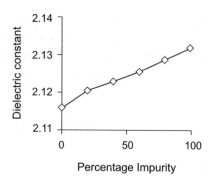

Fig. 1 Variation in dielectric constant with percentage impurity at optical frequency.

The dielectric constant increases with increase in the percentage of impurity and for pure vanaspati ghee, it approaches to 2.131. The specified range for consumption of this sample may be obtained when the amount of vanaspati ghee present in the sample is less than 8%. These variations are shown in Fig. 1.

The experimental values of dielectric constant measured at 9.46 GHz microwave frequency for the same samples varies from 2.463 (for pure desi ghee) to 2.532 (for pure vanaspati ghee). These variations are shown in Fig. 2. Corresponding to acceptable value of ε_∞ the dielectric constant of the sample at microwave frequency should not be more than 2.467. If it is more than 2.467, the sample may be immediately rejected. The dielectric loss and relaxation time measurements are also carried out and it is found that dielectric loss and relaxation time values for acceptable samples should not be more than 0.078 and 5.33×10^{-13} s, respectively.

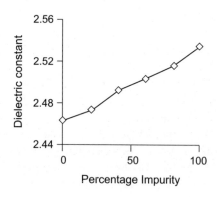

Fig. 2 Variation in dielectric constant with percentage impurity at microwave frequency.

The dielectric constant measured at a static frequency (300 KHz) for the same sample varies from 3.052 (for pure desi ghee) to 3.265 (for pure vanaspati ghee). These variations are shown in Fig. 3. It indicates that the acceptable value of dielectric constant should not be more than 3.071, otherwise sample should be rejected.

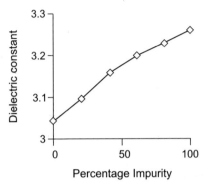

Fig. 3   Variation in dielectric constant with percentage impurity at static frequency.

More work is in progress and will be presented in near future. The proposed technique is an advanced technique and may be applied to check the purity of any liquid samples like checking the presence of arzimon in mustard oil, fluoride content in water, impurities in cold drinks etc.

Acknowledgements

Authors are thankful to AGMARK, Jaipur to use their available facilities for the present work and for some fruitful discussions on the subject. One of us (SA) is thankful to DST, Jaipur for providing financial support to carry out this work.

References

1. R.A. Jangid, D. Bhatnagar and J.M. Gandhi, "Study of dielectric behaviour of ternary mixtures of anilines at microwave frequency at room temperature 28°C" *Ind J Pure & Appl Phys*, **33** (1995) 135.

2. J.S. Yadav and J.M. Gandhi "Simple microwave technique for measuring the dielectric parameters of solids and their powders", *Ind J Pure & Appl Phys,* **30** (1992) 427.

3. H.J. Fischbeck and K.H. Fischbeck "Formula, Facts and Constants". Berlin: Springer-Verlag, 254 (1987).

Topics in Electromagnetic Waves: Devices, Effects and Applications
Edited by J. Behari
Copyright © 2005, Anamaya Publishers, New Delhi, India

14. Microwave Dielectric Constant of Soils

A.D. Vyas and D.H. Gadani

School of Sciences, Gujarat University, Ahmedabad, India

Abstract: Dielectric properties of six soil samples collected from various parts of Gujarat are measured at an X-band microwave frequency. The permittivity and dielectric loss of soils increase with moisture content in the soil. For sandy soil, an anomalous behaviour is observed, particularly for more than 80% sand content. The measured values of complex permittivity with moisture content were compared with those calculated by Hallikainen et al. model and Wang and Schmugge model. It is found that estimated values of ε' for different moisture content are in good agreement with the observed values of ε'. The estimated value of emissivity calculated from the measured values of complex permittivity decreases with the increase in moisture content in the soil.

Introduction

Microwave remote sensing technique to estimate the moisture content in the soil has gained considerable attention, because microwave sensors can operate in all weather conditions [1]. Further, microwave can penetrate deep in the soil in comparison to visible and infrared radiations. The technique involves either measurement of emissivity of soils using radiometers, or back scattering coefficient using an active sensor. Both the techniques depend on the dielectric constant of the soils, which is considerably affected by the moisture content in the soil. In order to interpret the data obtained by remote sensors, the study of variation of complex permittivity of the soils with moisture content is required. Considerable work has been done in this area in different countries [1]. India, being a large country, an attempt has been made to study the dielectric properties of soils of different regions of India (with moisture content) by some workers [2-7].

A comprehensive study of complex permittivity of soils of Gujarat State has been undertaken. Soil samples were collected from different regions of Gujarat and measurements of complex permittivity (laboratory conditions) were done for dry and wet soils at an X-band microwave frequency, as reported in our earlier papers. In continuation of this we report here complex permittivity of soil of Jamnagar district with moisture content. The earlier reported data on dielectric properties of other regions have also been included for comparison.

Empirical models [8-9] and mixing formula [10-11] have been developed for estimation of complex permittivity of soil water mixture. The determined values of ε^* of soils were also compared with the calculated values from the empirical models [8-9], and these results are also presented in this article.

Sample Collection

The samples of soils were collected from different regions of Gujarat (Saurashtra, South Gujarat, Sabarmati River, North Gujarat and Central Gujarat) in polythene bags and brought to the laboratory for measurement.

Experimental Procedure

Texture size was deteremined by mechanical fractionation and sedimentation technique and is listed in Table 1. Samples were dried in an oven at 110 °C for 24 h. Different levels of moisture were determined by adding distilled water to dry samples and allowing it to stand for 24 h to aid setting.

Table 1.   Textural composition and physical parameters of soils

Location	Soil texture(%)			Soil type	Wilting	Transition
(Region)	Sand	Silt	Clay		point (WP) cm^3/cm^3	moisture (W_t) cm^3/cm^3
Sabarmati River (Ahmedabad)	93	6.2	0.8	Sand	0.012	0.1708
Gandhinagar District	65	31	4	Sandy loam	0.045	0.1872
Amreli Dist.	11	78	11	Silt loam	0.118	0.2228
Valsad Dist.	7	62	31	Silty clay loam	0.211	0.2686
Palanpur District	82	16	1	Sand	0.02114	0.1698
Jamnagar District	12	50	38	Silty clay loam	0.2417	0.2834

The gravimetric moisture content was found by the relation
Percent moisture content

$$W_m = \frac{(\text{Weight of the wet soil} - \text{Weight of the dry soil})}{(\text{Weight of the dry soil})} \times 100\%$$

Hence the volumetric moisture content in the soil sample is calculated as

$$W_v = W_m \times (\text{Bulk density of the dry soil sample})$$

The wilting point and transition moisture of soil in terms of volumetric water content (cm^3/cm^3) were calculated using the Wang and Schmugge model as

$$WP = 0.06774 - 0.00064 \times \text{Sand} + 0.00478 \times \text{Clay}$$

where Sand and Clay are the sand and clay contents in percent of dry weight of the soil.

The transition moisture calculated as

$$W_t = 0.49\,WP + 0.165$$

are listed in Table 1.

At microwave frequencies the slotted line technique is generally used because it is convenient and readily available. The accuracy of measurements of dielectric constant depends on the accuracy with which the VSWR and the position of the voltage minimum can be found.

We used the two-point method for measurement of dielectric constant involving the solution of a complex transcendental equation, which is suitable for lossless and medium loss dielectric [12].

The experimental set up for the two-point method at X-band microwave frequency is shown in Fig. 1. First, with no dielectric in the short-circuited line, the position of the first minimum D_R in the slotted line was measured. Now the soil sample of certain length l_ε having certain moisture content was placed in the sample holder, such that the sample touches the short-circuited end. Now the position of the first minimum D on the slotted line and the corresponding VSWR, r were measured. This procedure was repeated for another soil sample of same moisture content for another soil sample length $l_{\varepsilon'}$. Now the propagation constant (in the empty waveguide) is calculated as

$$k = \frac{2\pi}{\lambda_g} \tag{1}$$

where $\lambda_g = 2 \times$ (Distance between successive mimima with empty short circuited waveguide).

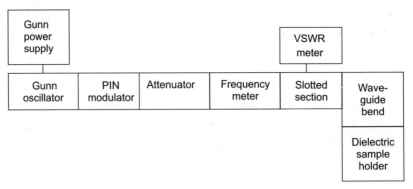

Fig. 1 Experimental set up for the two-point method at X-band microwave frequency.

The complex number $C\angle - \Psi$ is calculated using the equation

$$C\angle - \Psi = \frac{1}{jkl_\varepsilon} * \frac{1 - |\Gamma| * e^{j\phi}}{1 + |\Gamma| * e^{j\phi}} \tag{2}$$

where $\qquad \phi = 2k*(D - D_R - l_\varepsilon)$ $\qquad\qquad$ (3)

and $\qquad |\Gamma| = \dfrac{r-1}{r+1}$ $\qquad\qquad$ (4)

The solution of the complex transcendental equation

$$C\angle - \Psi = \frac{\tanh(T\angle\tau)}{T\angle\tau} \tag{5}$$

was obtained [14] to get conductance G_E and susceptance S_E. The dielectric constant ε' and the dielectric loss ε'' of the soil sample are then calculated as

$$\varepsilon' = \frac{G_E + \left(\dfrac{\lambda_g}{2a}\right)^2}{1 + \left(\dfrac{\lambda_g}{2a}\right)^2} \tag{6}$$

and

$$\varepsilon'' = \frac{-S_E}{1 + \left(\dfrac{\lambda_g}{2a}\right)^2} \tag{7}$$

where a is the width of the waveguide.

Results and Discussion

A plot of ε' and ε'' against volumetric soil moisture content for the soil of Jamnagar district is shown in Fig. 2. The real part of complex permittivity ε' increases gradually with the moisture content upto the transition moisture, after which it increases sharply with the moisture content in the soil. The imaginary part of the complex dielectric constant ε'' increases linearly with the moisture content in the soil and does not show a sharp increase after transition moisture. Similar behaviour was observed in our earlier studies for soils of different regions of Gujarat. This is due to the fact that in above transition moisture there are more free water molecules in comparison to bound water molecules. The bound water has low dielectric constant in comparison to free water. A comparison of ε^* of Jamnagar district soil with the soils collected from other regions of Gujarat is shown in Fig. 3. The value of ε' for dry soil is higher than those collected from other regions of Gujarat. This may be due to the presence of higher mineral content in this soil in comparison to other soils. It is evident from Fig. 3 that below transition moisture the data points (ε', ε'') for all six soils fall within a relatively narrow band, indicating that upto this moisture content the main parameter determining ε^* is the volumetric water content. After the transition moisture W_t, the variation of ε' and ε'' with moisture content is soil texture dependent. It is noticed that above transition moisture ε' value of soil is lower for samples having higher clay contents. Further, the ratio $\Delta\varepsilon'/\Delta W_v$ is higher for

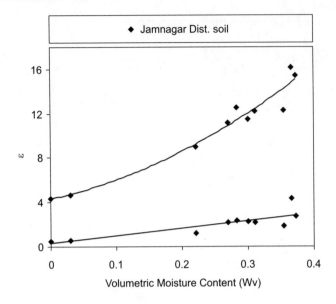

Fig. 2 Graph of complex permittivity of Jamnagar District soil for various moisture contents.

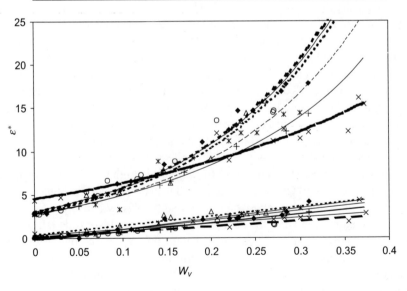

Fig. 3 Comparison of ε^* of Jamnagar District soil with the soils collected from other regions of Gujarat.

soil samples having higher sand content; this has been observed by previous investigators (Hallikainen et al, Wang and Schmugge, and Alex and Behari). It is due to a large specific surface area of clay particles in comparison to other basic components of soil, silt and sand. The large specific surface area of clay particles enables soil to retain greater moisture content in the form of bound water.

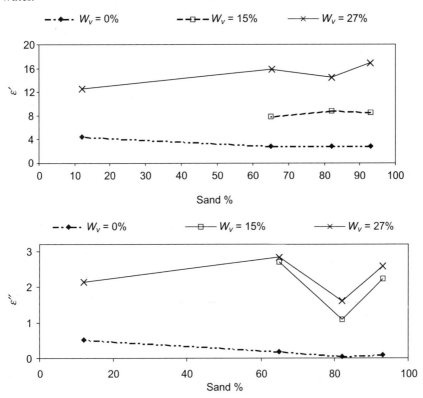

Fig. 4   Plot of ε' and ε'' against sand percentage.

A plot of ε' and ε'' against sand percent has been shown in Fig. 4, for dry, medium and high moisture contents. The value of ε' and ε'' increases with moisture content in the soil. Above 80% of sand content there is a sharp peak/ hump present, which cannot be explained with the available data. This anomalous behaviour of high sand content soil needs further attention, because, it may be exploited for soil characterization. Behari [13] observed similar behaviour in his study of frequency dependent dielectric behaviour of wet soil.

The complex permittivity of wet soils can be predicted, if the soil physical parameters and textures are known, by two prominent models given by Wang and Schmugge and Hallikainen et al. The complex permittivity of six soils of Gujarat with moisture content, calculated using both the models, are shown in Fig. 5 for sand, sandy loam and silty clay loam from Sabarmati, Gandhinagar and Valsad.

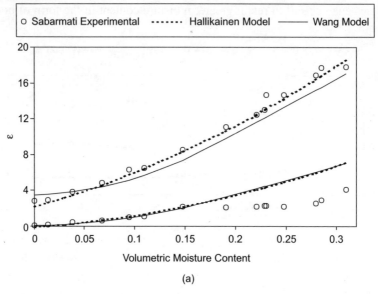

(a)

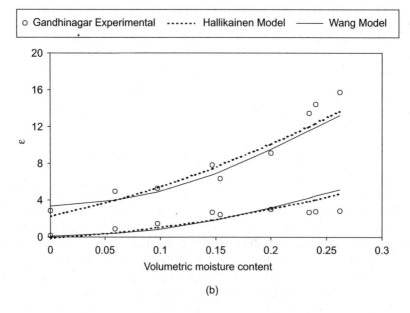

(b)

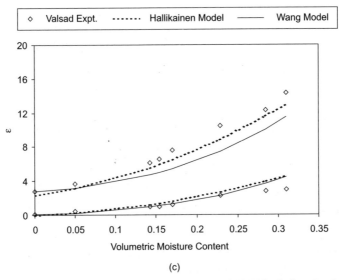

(c)

Fig. 5 (a) Sabarmati river sand, (b) sandy loam and (c) Valsad silty clay loam.

The experimentally observed values of ε^* have also been included in the figure for comparison. The ε' values with moisture content for sand and sandy loam soils are in agreement with those calculated from both the models. The ε' for various moisture content for silty clay loam soils are predicted well with Hallikainen et al. model, but Wang and Schmugge model predicts slightly less values of ε' with moisture content for these type of soils. The values of ε'' calculated from empirical models for all types of soils are in good agreement with the experimentally observed values upto 15% of moisture content in the soil, after which predicted ε'' values with moisture content are higher than the measured values of ε''.

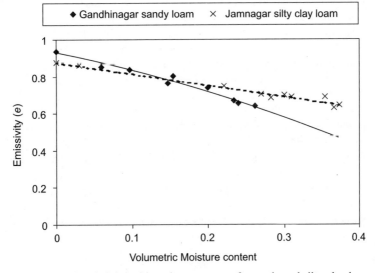

Fig. 6 Variation of emissivity with moisture content for sandy and silty clay loam soils.

The emissivity of soils from the measured values of complex permittivity for normal incidance can be calculated from the relation

$$e = 1 - \left|\frac{1-\sqrt{\varepsilon}}{1+\sqrt{\varepsilon}}\right|^2 \tag{8}$$

Fig. 6 shows the variation of emissivity calculated from the measured values of ε^* with moisture content for sandy loam and silty clay loam soils. It is evident from the figure that, for dry soils emissivity is less than one and decreases with increase in moisture content in the soil.

References

1. Microwave Remote Sensing (Active and Passive), F.T. Ulaby, R.K. Moore and A.K.Fung, vol. I, II, III. Artech House Inc. 1981,1982,1986.
2. Vyas, A.D., "Complex Permittivity of Sand and Sandy Loam Soils at Microwave Frequency," *Indian J. Radio & Space Physics*, **II**, pp. 169-170. August 1982.
3. Z.C. Alex and Behari J. "Laboratory evaluation of emissivity of soils", *Int. J. Remote Sensing*, **19**, No.7, pp.1335-1340, 1998.
4. Mishra U.S. and Behari J. "*In-situ* measurement of Dielectric Parameter of Soil at Microwave Frequencies," *Journal of the Indian Society of Remote Sensing*, **28,** No.1, 2000.
5. Pancholi K.C. and Khameshra, S.M "Complex dielectric permittivity of some Rajasthan soils at 7.114 GHz", *Indian J. of Radio and Space Physics*", **23,** June 1994, pp.201-204.
6. Shrivastava, S.K. "Transmission Line Model for Predicting the microwave emission from soil", Proceedings, National conference on Application of Remote Sensing, November 201, pp.1-7.
7. Calla, O.P.N, Borah, M.C., Vashishtha, P., Mishra, R., Bhattacharya, A. and Purohit, S.P. "Study of the Properties of dry and wet loamy sand soil at microwave frequencies", *Indian J. of Radio and Space Physics*, **28,** June 1999, pp.109-112.
8. Hallikainen, M.T., Ulaby, F.T., Dobson, M.C., El-Rays, M.A. and Lin-Kun Wu, "Microwave Dielectric Behaviour of wet Soil-part 1: Empirical Models and Experimental Observations," *IEEE Trans. Geosci. Remote Sensing,* **GE-23**, No.1, pp.25-33, January 1985.
9. Wang, J.R. and Schmugge, T.J. "An empirical model for the complex dielectric permitivity of soils as a function of water content," *IEEE Transactions on Geoscience and Remote Sensing,* **GE-18**, No.4, pp. 288-295, October 1980.
10. de Loor G.P., "Dielectric Properties of Heterogenious Mixtures containing Water", J. *Microwave Power,* **3**, pp. 67-73,1968.
11. Behari, J., Sahu, S.K. and Mishra, U.S. "Microwave Dielectric Constants of Soil", Physical Methods of Soil Characterization, Narosa, New Delhi, 2001, p. 25
12. Sucher, M. and Fox, J. 1963, Handbook of Microwave Measurements.

13. Behari, J. "Frequency Dependent Variation of Dielectric Parameters of Wet Soil", Microwave Measurement Technique and Applications, Anamaya Publishers, New Delhi, 2003, p. 71

14. Dielectric Materials and Applications, Edited by Arthur R. Von Hippel, 1954.

Topics in Electromagnetic Waves: Devices, Effects and Applications
Edited by J. Behari
Copyright © 2005, Anamaya Publishers, New Delhi, India

15. Microwave Dielectric Study of Dry and Water Saturated Samples of Various Grade Limestones and Sandstones

R.J. Sengwa and A. Soni

Microwave Research Laboratory, Department of Physics,
JNV University, Jodhpur-342 005, India

Abstract: Complex dielectric permittivity ($\varepsilon^* = \varepsilon' - j\varepsilon''$) of dry and water saturated 17 limestone samples i.e. cement grade, SMS grade, chemical grade and dolomite grade, and 14 different sandstone samples of Jodhpur region were carried out at 10.1 GHz and at room temperature. The percentage of chemical constituents of the sample composition, density and porosity of limestones and sandstones were used for the interpretation of their measured values of microwave permittivities. It is observed that their density governs the measured values of real part of dielectric constant ε' of equal percentage chemical composition limestones. But the variation in percentage of chemical composition significantly affected the ε' values of equal density dry limestones. In case of dry sandstones, ε' values of most of the samples were found density dependent. The density reduced permittivity ε'_{dr} of limestones and sandstones were also determined. It is found that the average ε'_{dr} value of limestones is 0.33 higher than average ε'_{dr} value of sandstones. The anomalous variation in ε'' values of dry limestone and sandstone samples confirm that no convincing correlation seems for all samples to correlate the ε'' values with the density and the chemical composition. The increase in permittivity values of water-saturated samples were found proportional to the percentage volume of water absorbed in the samples. The real part of dielectric constant of water saturated samples were theoretically computed by using mixing equation and also by using complex refractive index method (CRIM), and self-similar model (SSC). It is found that the practical ε' values of some samples are in good agreement with the ε' values determined from mixing equation but the samples for which mixing equation is not fitted properly, their measured ε' values were found close with the ε' values determined from CRIM equation. Further, the computed ε' values of dry limestones and sandstones from CRIM model and SSC model were found equal for the same sample. The microwave permittivities values were determined in view of their use in development of dielectric sensing technique for the exploration of limestone and sandstone deposition and estimation of water saturation level in ground water investigation using high frequency electromagnetic waves.

Introduction

The microwave dielectric study of dry and water saturated geologic materials helps in microwave remote sensing [1-2], planning ground penetrating radar (GPR) surveys [3-5], calibration of time domain reflectometry (TDR) measurements [6] and estimation of individual contribution and interaction contribution of the sample constituents to the microwave dielectric constant [7]. In rocks and sediments, real and imaginary parts of dielectric constant are primarily a function of mineralogy, density, porosity, water saturation, frequency and rock lithology and component geometries. Therefore, in heterogeneous geologic materials it is difficult to correlate the experimental dielectric values with all the petrological parameters and also with the chemical composition, simultaneously. Several investigators [3, 5-10] extensively studied the frequency dependent theoretical and experimental dielectric behaviour of dry and wetted sandstones. Literature survey shows that few attempts were also made to explore the microwave dielectric properties of limestone samples [2-4, 11-15]. So far no attempt was made to explore the microwave dielectric behaviour of large scale deposited various grade limestones and high quartz sandstones of Jodhpur region.

In the present study, an attempt is made to determine the experimental x-band microwave dielectric values of various grade limestones and sandstones of Jodhpur region in their dry state and also in water saturated state. The Jodhpur region limestones are extensively quarried for their use in black and white cement industries, in chemical industries and also in steel industries. Sandstones are quarried for their use as building stone from very old age. The studied samples of different grain sizes and colours were collected from the running opencast mines of Jodhpur region and are given in Tables 1 and 2 with their locations and chemical composition.

Table 1. Physical parameters and chemical composition of various grade limestone samples (in weight percentage)

Samples No.	Location	d (g/cc)	% ϕ	CaO	SiO$_2$	MgO	Al$_2$O$_3$	Fe$_2$O$_3$	AI	LOI
				Cement grade						
C1	Gotan	2.66	1.15	39.48	1.54	12.69	0.57	0.07	0.16	45.37
C2	Tankla-Khivser	2.58	0.46	48.72	7.66	1.20	0.81	0.07	0.42	40.10
C3	Gotan	2.74	0.09	43.96	2.52	7.45	0.35	0.09	0.44	44.01
C4	Gagwana	2.68	0.75	49.84	0.22	4.83	0.51	0.11	0.14	44.04
C5	Basni Harisingh	2.70	0.18	36.40	7.90	12.89	0.56	0.26	0.38	41.58
C6	Ras-Jetaran	2.66	0.18	36.40	15.84	0.60	5.59	0.57	11.88	28.35
				SMS grade						
S1	Gotan	2.39	4.34	53.76	0.40	0.30	0.29	0.03	0.68	43.66
S2	Gagwana	2.53	0.76	54.32	0.24	0.10	0.65	0.03	0.44	43.66
S3	Manakpur	2.56	0.32	54.04	0.38	0.20	0.66	0.04	0.48	43.75
S4	Deh	2.53	0.07	54.04	0.24	0.20	0.70	0.06	0.48	43.28
				Chemical grade						
CH1	Gotan	2.26	7.77	51.32	1.00	2.21	0.58	0.10	0.38	43.45
CH2	Jaisalmer	2.46	7.21	53.20	1.22	0.20	0.17	2.03	0.60	42.50
CH3	Jaisalmer	2.49	5.35	52.36	0.56	0.40	0.14	3.08	0.74	42.38
				Dolomite						
D1	Phalodi	2.37	7.85	31.64	0.26	20.15	0.54	0.18	0.54	46.26
D2	Basni Harisingh	2.78	0.35	31.64	0.56	19.94	0.57	0.23	0.34	46.04
				Miscellaneous						
M1	Deh	2.09	9.30	9.52	48.30	5.03	5.12	6.44	7.58	18.00
M2	Kota	2.63	0.89	39.76	19.18	0.20	1.58	1.30	5.88	32.00

Table 2.   Physical and petrophysical parameters of different samples of Jodhpur sandstones

Samples No.	Location	d (g/cc)	% ϕ	Grain size (mm)	SiO_2	Chert/ Volcanic/ Clatic sand	Al_2O_3/CaO MgO clay (illite)	Fe_2O_3
1	Fidusar	2.43	4.16	0.25-0.13	98	–	1	1
2	Machia Safari Park	2.35	5.31	2.00-0.25	96	–	1	3
3	Machia Safari Park	2.30	5.79	0.25-0.06	96	–	2	2
4	Fidusar	2.45	3.48	0.25-0.13	98	–	1	1
5	Fidusar	2.44	3.72	0.50-0.03	96	–	1	3
6	Masooria	2.34	6.05	0.50-0.06	95	–	3	2
7	Balesar	2.25	6.25	0.25-0.13	95	–	1	4
8	Bhopalgarh	2.28	10.10	0.13-0.06	95	–	1	4
9	Dholpur	2.32	5.89	0.25-0.06	98	–	1	1
10	Dholpur	2.31	6.41	0.25-0.06	98	–	1	1
11	Bola Ram Devri	2.31	3.07	0.60-0.03	92	1	2	5
12	Surpura	2.52	3.38	0.13-0.03	90	2	2	6
13	Ganesh Doongari	2.46	3.33	0.13-0.03	83	3	7	7
14	Chopasani	2.34	7.09	0.13-0.03	79	3	8	10

Experiment

For dielectric measurements, x-band wave-guide dimensions sections of two different thickness of each limestone and sandstone were prepared. The measurements of ε' and ε'' at 10.1 GHz were made with sample length variation method [16, 17] using a short-circuited slotted wave-guide operating in TE_{10} mode. Each sample was fully dried before dry sample measurements. For the dielectric study of water saturated samples, same samples were dipped in deionized water for three days for fully water saturation. The measured values of microwave complex permittivity are recorded in Tables 3 and 4.

Results and Discussion

Dielectric Behaviour of Dry Limestone Samples

The ε' values of dry limestone samples of Jodhpur region were found in the range of 5.55 to 7.81 at room temperature (sample M1 excluded). Table 1 shows that the constituents of the studied limestone samples are CaO, MgO, SiO_2,

Table 3. Values of real part of dielectric constant and loss of dry and water saturated various grade limestone samples at 10.1 GHz, and at room temperature

Samples No.	d (g/cc)	Dry samples			Water saturated samples				Theoretical ε' values		
		ε'	ε''	ε'_{dr}	% ϕ	ε'_s	ε''_s	$\varepsilon'_s - \varepsilon'$	Mixing equation	CRIM Model	SSC Model
Cement grade											
C1	2.66	6.90	0.01	2.07	1.15	7.39	0.05	0.49	7.55	7.22	7.21
C2	2.58	7.30	0.04	2.16	0.46	7.56	0.05	0.26	7.56	7.43	7.43
C3	2.74	7.27	0.02	2.06	0.09	7.49	0.06	0.22	7.32	7.30	7.29
C4	2.68	7.55	0.13	2.13	0.75	7.82	0.17	0.27	7.97	7.77	7.76
C5	2.70	7.50	0.04	2.11	0.18	7.51	0.07	0.01	7.61	7.55	7.55
C6	2.66	7.71	0.41	2.16	0.18	8.08	1.01	0.37	7.81	7.76	7.76
SMS grade											
S1	2.39	6.91	0.01	2.25	4.34	8.31	0.50	1.40	9.34	8.17	8.14
S2	2.53	7.37	0.05	2.20	0.76	7.71	0.21	0.34	7.79	7.59	7.58
S3	2.56	7.58	0.02	2.21	0.32	7.75	0.04	0.17	7.76	7.67	7.67
S4	2.53	7.81	0.03	2.25	0.07	7.96	0.04	0.15	7.85	7.83	7.83
Chemical grade											
CH1	2.26	6.12	0.03	2.23	7.77	10.78	3.55	4.66	10.54	8.40	8.38
CH2	2.46	6.65	0.35	2.16	7.21	9.60	1.10	2.95	10.71	8.79	8.75
CH3	2.49	7.14	0.51	2.20	5.35	8.81	0.33	1.67	10.13	8.73	8.68
Dolomite											
D1	2.37	5.55	0.11	2.06	7.85	8.21	0.20	2.66	10.06	7.81	7.80
D2	2.78	7.04	0.01	2.02	0.35	7.16	0.03	0.12	7.24	7.14	7.14
Miscellaneous											
M1	2.09	4.56	0.63	2.07	9.30	12.27	4.36	7.71	6.28	7.15	7.20
M2	2.63	6.62	0.70	2.05	0.89	7.03	1.17	0.41	7.12	6.87	6.86

Table 4. Values of real part of dielectric constant and loss of dry and water saturated different sandstone samples at 10.1 GHz, and at room temperature

Samples No.	d (g/cc)	Dry samples			Water saturated samples				Theoretical ε' values		
		ε'	ε''	ε'_{dr}	% ϕ	ε'_s	ε''_s	$\varepsilon'_s - \varepsilon'$	Mixing equation	CRIM Model	SSC Model
1	2.43	4.09	0.007	1.79	4.16	5.57	0.56	1.48	6.54	5.15	5.16
2	2.35	3.77	0.021	1.76	5.31	5.34	0.59	1.57	6.92	5.11	5.13
3	2.30	4.01	0.007	1.83	5.79	7.52	3.05	3.51	7.43	5.50	5.53
4	2.45	4.20	0.002	1.80	3.48	5.20	0.34	1.00	6.24	5.08	5.09
5	2.44	4.06	0.002	1.78	3.72	5.25	0.27	1.19	6.25	5.00	5.00
6	2.34	4.16	0.027	1.84	6.05	7.03	2.23	2.87	7.71	5.74	5.76
7	2.25	3.48	0.514	1.74	6.25	5.38	0.55	1.90	7.20	5.04	5.09
8	2.28	4.03	0.068	1.84	10.10	10.01	2.63	5.98	9.98	6.79	6.88
9	2.32	3.91	0.001	1.80	5.89	5.49	1.44	1.58	7.39	5.42	5.45
10	2.31	3.97	0.001	1.82	6.41	5.51	0.28	1.54	7.75	5.63	5.67
11	2.31	4.00	0.007	1.82	3.07	4.98	0.42	0.98	5.81	4.76	4.77
12	2.52	4.84	0.102	1.87	3.38	6.80	1.77	1.96	6.81	5.73	5.72
13	2.46	4.62	0.447	1.86	3.33	6.45	1.47	1.83	6.57	5.49	5.48
14	2.34	3.89	0.444	1.79	7.09	6.13	1.16	2.24	8.08	5.74	5.78

Al_2O_3 and Fe_2O_3. Earlier [7] measured values of ε' and ε'' of these oxides at 10 GHz are recorded in Table 5. The major constituent of limestone is CaO. The increase in weight percent of SiO_2 and MgO will reduce the ε' value, whereas the increase in weight percent of Al_2O_3 and Fe_2O_3 in limestone composition will increase the ε' value of limestone due to their higher dielectric constants (Table 5). From microwave dielectric study of various rocks and minerals [18, 19], it is already confirmed that the variance in ε' values of dry geologic materials is mainly governed by the sample bulk density and their chemical composition. Olhoeft and Strangway [20], in a lunar sample study, also confirmed that density is the primary control on high-frequency permittivities of dry specimens. The density reduced permittivity ε'_{dr} of dry limestones and sandstones were determined using

$$\varepsilon'_{dr} = (\varepsilon')^{1/d} \tag{1}$$

where ε' is the measured permittivity and d the dry sample bulk density. The computed values of ε'_{dr} for the studied limestones and sandstones are recorded in Tables 3 and 4, respectively.

Table 5.   Values of ε' and ε'' of different oxides of limestone at 10 GHz [7]

Oxides	CaO	SiO_2	MgO	Al_2O_3	Fe_2O_3
$\varepsilon'-j\varepsilon''$	8.22–j 0.12	4.43–j 0.04	5.03–j 0.17	12.66–j 1.31	16.58–j 0.93

In case of limestone samples the average value of ε'_{dr} is found 2.14 ± 0.14. Earlier [20] for moon rocks the value of ε'_{dr} is found 1.93 ± 0.17, whereas for different earth rocks [18] it observed that the value of ε'_{dr} is 1.96 ± 0.14. In case of various grade limestone samples of Jodhpur region the observed value of ε'_{dr} (Table 3) is found higher in comparison to earlier [18, 20] reported ε'_{dr} values.

In case of cement grade samples, the variation in their density is small and due to anomalous variation in the samples chemical constituents, no convincing correlation is observed between ε' values and their density and chemical composition. In case of SMS grade samples, it seems that the increase in sample bulk density and percentage of $Fe_2O_3+Al_2O_3$ in very small quantity increases the ε' values from samples S1 to S4. Simultaneously, the decrease in percentage of low ε' values constituents, i.e. SiO_2+MgO also contributed in increase in the ε' values (Table 3).

In chemical grade samples there is significant increase in percentage of $Fe_2O_3+Al_2O_3$ and simultaneously decrease in percentage of SiO_2+MgO and, hence, the observed ε' values increases from CH1 to CH3. Further the increase in density also contributed in the increase in ε' values of chemical grade samples. In case of dolomite samples it is clear that the increase in ε' value of D2 in comparison to D1 is only due to the higher density of D2 because both samples have equal chemical composition. The variation in ε' values of marli limestone M1 and Kota limestone M2 are also according to the change in density and their chemical composition. In case of dolomite and miscellaneous limestone samples, there is large difference in their density but the values of ε'_{dr} (Table 3) were found nearly equal, which confirms that the permittivity of these dry samples is mainly governed by their bulk density. Further the observed values of ε' of Jodhpur region limestones were found in good agreement with the earlier reported ε' values of other regions limestone samples [2, 3, 11-15].

Only in case of chemical grade sample the increase in ε'' values is found according to the increase in both the percentage of Fe_2O_3 and sample bulk density. But other samples show the anomalous variation in ε'' values with increase in density.

Dielectric Behaviour of Water Saturated Limestone Samples

From the comparative enhance in real part of dielectric constant ε'_s values of water saturated limestone samples of chemical grade, SMS grade, dolomite and miscellaneous, it is found that enhance in permittivity $\varepsilon'_s - \varepsilon'$ due to water saturation, follows the value of samples porosity ϕ (Table 3). But in case of cement grade samples the enhance in ε' value due to water saturation is not found according to their increase in porosity. These cement grade samples show the anomalous variation in the values of ε'_s with increase in their porosity.

The real part of dielectric constant of water saturated limestones ε'_s were theoretically computed by using the mixing equation

$$\varepsilon'_s = \phi\,\varepsilon'_w + (1-\phi)\,\varepsilon' \tag{2}$$

by using complex refractive index method (CRIM)

$$\sqrt{\varepsilon'_s} = \phi\sqrt{\varepsilon'_w}\,(1-\phi)\sqrt{\varepsilon'} \tag{3}$$

and also by using self similar model (SSC) [21]

$$\left(\frac{\varepsilon'-\varepsilon'_s}{\varepsilon'-\varepsilon'_w}\right)\left(\frac{\varepsilon'_w}{\varepsilon'_s}\right)^{1/3} = \phi \tag{4}$$

where ε'_w and ε' are the dielectric permittivity values of water and dry samples at 10.1 GHz respectively, and ϕ is the sample porosity. The value of $\varepsilon'_w = 63$ is used [22] in the above equations.

Table 3 shows that the computed ε'_s values of all cement grade samples and S2, S4, CH1, D2 and M2, using mixing equation (1) are in good agreement with the experimental values. But for samples S1, CH3 and D1, experimental values of ε'_s were found close to the ε'_s values computed from CRIM model. Further it is found that the experimental ε'_s values of CH2 and M1 do not match with theoretical values obtained from equations (1) and (2). For these samples some more modified mixing equation is required. Table 3 also shows that the computed values of ε'_s using CRIM model and SSC model were found equal for same sample.

Table 3 also shows that enhance in the values of ε''_s of most of the water saturated limestones in different grades is according to the increase in their porosity. But the enhance in ε''_s in different grade samples is not same, which may be due to the variation in samples pore geometries and also the different kind of interactions of water molecules with the sample constituents.

Dielectric Behaviour of Dry Sandstone Samples

The ε'_s values of Jodhpur region dry sandstone samples and two samples of Dholpur sandstones were found in the range of 3.5 to 4.84 with an average value 4.07 at 10.1 GHz (Table 4). These values are higher than the $\varepsilon' = 3.45$ value of Berea sandstone at 1 GHz [13]. But the average ε' value of Jodhpur sandstones were found slightly lower than $\varepsilon' = 4.43$ value of bulk crystalline quartz at 10 GHz [7]. Further the ε' values of Jodhpur sandstones were found of the order of the ε' value of calcarious sandstone [23] of Jodhpur region. All the studied Jodhpur sandstones have very high percentage of SiO_2 and the variation in their bulk density is small (Table 2). Therefore, it is difficult to correlate the ε' values of all these sandstone samples with their bulk density and chemical composition, simultaneously. But from Table 4 it seems that the ε' values of some sandstone samples can be correlated with their density. The computed values of density-reduced permittivity ε'_{dr} of these sandstones are recorded in Table 4. The average values of ε'_{dr} of Jodhpur sandstones were found 1.81 ± 0.07. This value is 0.33 lower than the value of Jodhpur limestones. The average of ε'_{dr} values of limestones and sandstones is 1.98, which is in good agreement with the ε'_{dr} values reported earlier [18] for various earth rocks.

The ε'' values of dry sandstone samples at 10.1 GHz were found in the range from 0.001 to 0.5 (Table 4). There is no direct convincing correlation between ε'' values and bulk density of dry samples.

Dielectric Behaviour of Water Saturated Sandstone Samples

Most of the studied water saturated Jodhpur sandstone samples show that the enhance in their real part of dielectric constant is proportional to the water absorbed in the samples i.e. proportional to their porosity. The theoretical computed ε'_s values of samples 3, 7, 8, 12 and 13 using mixing equation are in good agreement with the experimental observed ε'_s values. But in case of sandstone samples 1, 2, 4, 5, 7, 9, 10, 11 and 14, the experimental evaluated ε'_s values were found in good agreement with the ε'_s values computed from CRIM model (Table 4). In case of Jodhpur sandstones it is also found that the ε'_s values evaluated by CRIM model are equal to the ε'_s values computed from SSC model of a sample. This confirms that the CRIM model and SSC model have same effect on microwave dielectric permittivity values of water saturated Jodhpur limestones and sandstone samples.

The ε''_s values of water-saturated sandstones were found much higher than ε'' values of dry samples. Some samples show large increase in ε''_s values. The anomalous variation in ε''_s values is observed with the increase in sample porosity.

Conclusions

From the present study, the effect of samples bulk density, chemical composition and water saturation on the microwave dielectric constant of various grade limestones and sandstones of Jodhpur region has been characterized. The sample bulk density governs the ε' values of dry limestone and sandstone samples but chemical comoposition also influences the real part of microwave dielectric constant. For water-saturated samples, porosity of the sample plays the major role in enhance in the values of dielectric constants. Further the laboratory measured precise permittivity values of dry samples of limestones and sandstones and their porosity dependence real part of dielectric constant with water saturation will help in the calibration of the dielectric transducer in microwave dielectric sensing technique for estimation of water level in ground water investigation of these rocks bearing areas.

Acknowledgements

The authors acknowledge the financial support from the Department of Science and Technology, Government of Rajasthan, Jaipur. One of the authors (A. S.) is thankful to UGC, New Delhi for providing J.R.F.

References

1. F.T. Ulaby, R.K. Moore and A.K. Fung, "Microwave remote sensing: active and passive (Vol. III) – From theory to applications" Dedham, Massachusetts, 1986.
2. B. Cervelle and X. Jin-Kai, "Dielectric properties of minerals and rocks: Applications to microwave remote sensing". In: Advanced Mineralogy (Vol. I), A. S. Marfunin (Eds.), Springer-Verlag, Berlin, 1994.
3. A. Martinez and A. P. Byrnes, "Modeling dielectric-constant values of geologic materials: An aid to ground-penetrating radar data collection and interpretation", Current Res. *Earth Sci., Kansas Geological Survey Bulletin*, **247.** Part I, 1–16, 2001.
4. G.A. McMechan, R. G. Loucks, X. Zeng and Mescher P, "Ground penetrating radar imaging of a collapsed paleocave system in the Ellenburger dolomite-central Texas", *J. Appl. Geophys.,* **39,** 1–10, 1998.
5. L.J. West, K. Handley, Y. Huang and M. Pokar, " Radar frequency dielectric dispersion in sandstone: Implications for determination of moisture and clay content", *Water Resources Res.*, **39,** 1026–1032, 2003.
6. A.A. Ponizovsky, S.M. Chudinova and Y.A. Pachepsky, "Performance of TDR calibration models as affected by soil texture", *J. Hydrology,* **218,** 35–43, 1999.
7. S. Sharif, "Chemical and mineral composition of dust and its effect on the dielectric constant", *IEEE Trans. GeoSci. Remote Sensing*, **33,** 353–359, 1995.
8. R. Knight and A. Endres, "A new concept in modeling the dielectric response of sandstones: Defining a wetted rock and bulk water system", *Geophysics*, **55,** 586–594, 1990.
9. S. Capaccioli, M. Lucchesi, R. Casalini, P.A. Rolla and N. Bona, "Influence of the wettability on the electrical response of microporous systems", *J. Phys. D: Appl. Phys.,* **33,** 1036–1047, 2000.
10. J. Ph. Poley, J.J. Nooteboom and P.J. Wall, "Use of VHF dielectric measurements for borehole formation analysis", *Log Analyst,* **19,** 8–30, 1978.

11. S.O. Nelson, "Determining dielectric properties of coal and limestone by measurement on pulverized samples", *J. Microwave Power,* **31**, 215–220, 1996.

12. S.O. Nelson, and P. G. Bartley, "Open ended coaxial-line permittivity measurements on pulverized materials", *IEEE Trans. Inst. Meas.,* **47**, 133–137, 1998.

13. R. Freedman and J.P. Vogiatzis, "Theory of microwave dielectric constant logging using the electromagnetic wave propagation method", *Geophysics,* **44**, 969–986, 1979.

14. G.V. Keller, "Rock and mineral properties" in M.N. Nabighian Eds., Electromagnetic methods in applied Geophysics", *Soc. Expl. Geophys.,* 13–51, 1987.

15. R.J. Sengwa and A. Soni, "Microwave x-band dielectric properties of selected limestone samples", Proc. Nat. Conf. Microwaves, Antennas & Propagation, Jaipur, 349–351, 2001.

16. W.B. Westphal, "Dielectric measuring techniques in A.R. Von Hipple Eds. Dielectric materials and applications", Wiley, New York 1954.

17. S.O. Nelson, "A system for measuring dielectric properties at frequencies from 8.2 to 12.4 GHz", *Trans. ASAE,* **15**, 1094–1098, 1972.

18. F. T. Ulaby, T. H. Bengal, M.C. Dobson, J.R. East, J.B. Garvin and D.L. Evans, "Microwave dielectric properties of dry rocks", *IEEE Trans. Geosci. Remote Sensing,* **28**, 325–336, 1990.

19. S.O. Nelson, D.P. Lindroth and R.L. Blake, "Dielectric properties of selected minerals at 1 to 22 GHz", *Geophysics,* **54**, 1344–1349, 1989.

20. G.R. Olhoeft and D. W. Strangway, "Dielectric properties of the first 100 meters of the moon", *Earth Planet Sci. Lett.,* **24**, 394–404, 1975.

21. P. N. Sen, C. Scala and M. H. Cohen, "A self-similar model for sedimentary rocks with applications to dielectric constants of fused glass beads", *Geophysics,* **46**, 781–795, 1981.

22. R.J. Sengwa and K. Kaur, "Dielectric dispersion studies of poly(vinyl alcohol) in aqueous solutions", *Polym. Int.,* **49**, 1314–1320, 2000.

23. R.J. Sengwa, A. Soni and B. Ram, " Dielectric behaviour of shales and calcareous sandstone of Jodhpur region", *Indian J. Radio & Space Phys.,* **33**, 329-335, 2004.

Topics in Electromagnetic Waves: Devices, Effects and Applications
Edited by J. Behari
Copyright © 2005, Anamaya Publishers, New Delhi, India

16. Time Domain Reflectometry and Ground-Penetrating Radar Techniques for *in situ* Soil Moisture Measurement

J. Behari

School of Environmental Sciences,
Jawaharlal Nehru University, New Delhi-110067, India

Abstract: The development and continuing refinement of new techniques have significantly enhanced our ability to monitor the storage and movement of soil water *in situ*. However, no single approach serves the purpose that has the best overall performance for a range of soil, crop and landscape conditions. Time domain reflectometry, which is based on dielectric properties of soil and provide point measurement, is suitable for automatic, precise, rapid and reliable measurement of soil water. Ground-penetrating radar, on the other hand, offers a fast and non-destructive way for estimating dielectric constant and is suitable for mapping of large areas. This is sensitive to subsurface fluid flow process and appears promising for such applications. The aim of this article is to discuss these soil water measuring techniques.

Introduction

The restrictive use of neutron probe, the rapid advancement and the decreasing cost of the non-nuclear methods in recent years, brought about to compare these methodologies. This also defines decision-making process for assessing the characteristics of technology in relation to work objectives. However, soil water measurements encounter particular problems related to the physics of the method used.

Soil water is a highly dynamic entity, exhibiting substantial variation in both time and space. This is particularly true near the soil surface, and in the presence of active plant roots (Or and Wraith, 2000). Soil water is also the means for solute transport, including nutrients and soil contaminants. Accurate measurement of soil water is hence crucial for the better management of irrigation water and rainfall capture. Crop yields are generally more closely related to soil water availability than to any other soil and meteorological variable. Therefore, the effective use of soil water data requires more frequent and accurate measurements and the technique should be rapid, reliable, simple, cost effective and also preferably non-destructive. These data are a valuable part of agricultural, environmental and ecological research. Since most of the data acquisition are conducted on multiple sites, the accessibility of these is an important consideration when selecting an automated system for measuring soil water content (Veldkamp and O'Brien, 2000). Great effort has been devoted in the last decades to the development of new soil water-content sensors based on TDR and Ground Penetrating Radar (GPR). Presently we will describe these two techniques.

Time Domain Reflectometry (TDR)

Working Principles

TDR is a method of determining soil water content by making use of the fact that the dielectric constant of water is much higher than that of the other soil constituents. The method involves measuring the propagation velocity of an electromagnetic pulse travelling along parallel metallic probes (rods) embedded in the soil. This measurement later converted to the volumetric moisture content of soil by various models (Topp et al., 1980, Dalton et al., 1984, Dalton and van Genuchten, 1986).

The permittivity ε is a complex quantity, but for soil water content measurements, the imaginary part can be neglected. This part of complex permittivity represents energy absorption by the soil as a result of dielectric losses. The permittivity is then equal to the real part in magnitude. Thus, for vacuum and air $\varepsilon =1$, for water $\varepsilon \approx 81$, while for most water mixed with soil mineral soil components $\varepsilon \approx 3$ to 7 (Zegelin et al., 1992). As a result, the permittivity of moist soil varies strongly with water content and can be used to determine soil water content.

Frequency Domain Reflectometry (FDR) is also developed for continuous measurement of soil water content. This system uses the dielectric properties of water but in a different manner than TDR (Bilskie, 1997). In contrast to TDR, FDR sensor contains an Application Specific Integrated Circuit (ASIC) which measures real and imaginary part of the complex dielectric permittivity simultaneously by the sensor rods at the single frequency of 20 MHz. The ASIC increases the accuracy of the measurements and eliminates influences of lengths of cables, quality of cables, connectors, and switches, making multiplexing easier and cheaper (Dirksen and Hilhorst, 1995).

TDR operates in the frequency range of 1 MHz to 1 GHz, well below the relaxation frequency of water. Hoekstra and Delaney (1974) and Davis and Annan (1977) reported little frequency dependence of ε' across this range, though the electrical conductivity contributes to dielectric loss if the solution contain ions (De Loor, 1968). ε'' is generally small and insignificant in non-saline homogenous soils.

The use of remote shorting diodes and calibrated reference airlines can, in many cases, considerably improve the accuracy of TDR measurements (Hook et al. 1992; Topp et al. 1996). The signal to noise ratio of the reflected signals can be increased by using remotely switched diodes. This combined with a waveform subtraction procedure, provide reliable identification of the two reflections that define ε_{air}. The high resolution TDR system has an excellent advantage of detecting very small changes in soil water content. The system can, thus, be used for quantifying the effects of temperature variations, on the apparent dielectric constant of soils with different water contents (Pepin et al. 1995).

In TDR the dielectric constant of soil is determined by measuring the transition time t of a high frequency electromagnetic pulse (e.g. of the orders of 140 ps) launched along a pair of parallel wave guides of known length L buried inside the soil medium. As at the end of wave guide the launched electromagnetic pulse is reflected back to its source, so the path length of the voltage pulse is twice the length of waveguide. Thus, the propagation velocity of pulse V, can be written as

$$V = 2L/t \tag{1}$$

The time delay provides information about the real part of the dielectric constant.

In the electrical transmission line theory the electromagnetic wave propagation velocity in a transmission line is given by (Marshall and Holmes, 1988)

$$V = C \left[\frac{1}{2} \varepsilon \{1 + (1 + \tan^2 \delta)^{\frac{1}{2}}\} \right]^{-\frac{1}{2}} \tag{2}$$

$\tan \delta = \{\varepsilon'' + (\sigma_{dc}/\omega\varepsilon_o)\}/\varepsilon'$, where σ_{dc} is the zero frequency electrical conductivity.

At very high frequency $\delta \to 0$, so that

$$V = C/(\varepsilon_{ap})^{1/2} \tag{3}$$

where C is the velocity of an electromagnetic wave in free space or vacuum ($C = 3 \times 10^8$ ms^{-1}) and ε_{ap} the apparent dielectric constant of the soil being measured. From (1) and (3), we get

$$\varepsilon_{ap} = \left[\frac{ct}{2L} \right]^2 \tag{4}$$

Putting $L = 3.0.0$ cm, $\varepsilon_{ap} = 25.0$ and $C = 3.0 \times 10^8$ m s^{-1}, the time of propagation turns out to be 5 ns.

The apparent dielectric constant ε_{ap} depends on soil moisture content θ, following a linear relationship of the type

$$\sqrt{\varepsilon_{ap}} = C_1\theta + C_2 \tag{5}$$

where C_1 and C_2 are constants which depend on the soil type (Topp et al., 1994; Ferre et al., 1996). This signifies the need of only two point calibration.

The advent of diode-shorting techniques in TDR soil water content probes (Hook et al., 1992; Ferre et al. 1996) has led to the design of one piece profile probes. In a typical design, 6 shorting diodes are placed along a transmission line, spaced 15 or 30 cm apart. The interval between each diode and its neighbor is called a segment. In six-diode probe, there are thus five segments. The diodes are configured to apply a short circuit across the transmission line when activated. Figures 1 and 2 illustrate the use of multiplexing probe. The probe can be implanted at various locations in the field by choosing a suitable cable length within a radius of 50-100 cm with respect to the body of the apparatus. The system has data logging facility to continuously monitor the soil moisture profile over a period of time.

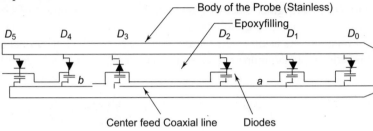

Fig. 1 Five segment, six-diode profile probe. A high frequency pulse propagated from the D_2 location spreads in both directions towards the ends.

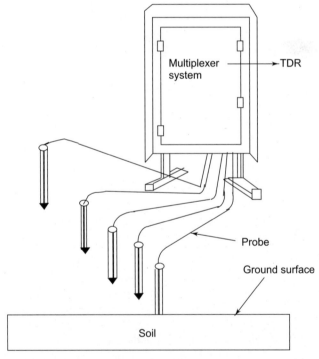

Fig. 2 The TDR multiplexing technology to measure soil moisture produces increased signal levels, enabling it to operate in a broader range of soils. These five probes are shown to be simultaneously operative.

Technically the round trip propagation time for a radio frequency (RF) pulse travelling along a particular segment is obtained by shorting the first diode in a segment, and measuring the time of arrival of the reflection from the discontinuity, introduced by the short circuit. This procedure is repeated for the second diode in the segment, and the two time intervals subtracted. The segment is calibrated by performing time interval measurement in dry sand (moisture content = 0) and bulk water (moisture content = 100%). The known dielectric constants of water and dry sand are used to calculate the coefficients of a linear equation relating time interval to water content. These coefficients are slightly different for each segment. This procedure amounts to performing a linear interpolation between two known points. There is an implicit assumption that time delay and soil water content are linearily related. The validity of this assumption has been demonstrated by overwhelming statistics obtained from measurements (Hook et al., 1992).

Water content measurements by TDR compare favourably with conventional and neutron scattering technique (Bandyopadhyay, 1995). TDR measurements average soil water content over a depth depending upon the wavelength. Wave guides upto 60 cm length have so far been used (Topp et al., 1994). Using TDR it is possible to infer soil electrical conductivity as well as soil water content (Dalton, 1992, Sundara Sarma et al., 1992). The effective use of TDR water content values for water balance monitoring depends on rapid and reproducible recovery of data from a number of representative locations. These requirements led to the development of automated analyses of the TDR trace (Zegelin et al., 1992) and multiplexing capabilities, which allow measurement of many locations using a single TDR unit.

It may be mentioned that the accuracy of the TDR method is limited by the resolution of the TDR system itself. Evaluation and improvements are called for in the existing TDR instruments for their effective use for soil water content measurement, particularly in the design of probe type. However its superiority persists owing to its versatility and easy maneuverability.

Ground Penetrating Radar (GPR)

GPR has been employed to follow the wetting front movement to monitor changes in soil moisture content (Vellidis et al., 1990). GPR is also suitable method for monitoring moisture content changes in the vadose zone and permit relatively large measurement scales, appropriate for hydrological models of unsaturated processes (Binley et al., 2001). GPR is a near-surface geophysical technique that can provide high resolution images of the dielectric properties of the top few tens of meters of the earth (Knight, 2001; van Dam and Schlager, 2000).

Measurements

GPR methods perform best in sites lacking highly electrically conductive materials, such as clay-rich soils (Davis and Annan, 1989). GPR wave attributes such as the electromagnetic (EM) wave velocity and attenuation are governed by electrical parameters including the electrical conductivity and the dielectric constant, both of which depend on water saturation (Daniels, 1996). For common earth materials (Davis and Annan, 1989), the EM wave velocity is related to the dielectric constant through the simple relationship.

$$v \approx \frac{C}{\sqrt{\varepsilon_{\text{eff}}}} \tag{6}$$

where ε_{eff} is the effective dielectric constant, which can be related to water saturation with a petrophysical model (Roth et al. 1990, Topp et al. 1980). The model used by Roth et al. (1990) gives the effective dielectric constant as

$$\varepsilon_{\text{eff}} = [(1-\varphi)\sqrt{\varepsilon_s} + S_w\varphi\sqrt{\varepsilon_W} + (1-S_w)\varphi\sqrt{\varepsilon_a}]^2 \tag{7}$$

where ε_s, ε_w and ε_a are the dielectric constants for the solid, water, and air components of the soil, respectively, φ is the soil porosity, and S_w is the water saturation.

In crosshole GPR applications, high-frequency EM pulses (commonly with central frequencies of 100 or 250 MHz) are propagated between boreholes in various antenna configurations (Fig. 3). A GPR wave attribute that is potentially sensitive to the distribution of water saturation is the arrival time. The combination of (6) and (7) gives the GPR inferred average water saturation at a given depth as

$$S_w, \text{GPR} = \frac{TcL^{-1} - (1-\varphi)\sqrt{\varepsilon_s} - \varphi\sqrt{\varepsilon_a}}{\varphi\left(\sqrt{\varepsilon_w} - \sqrt{\varepsilon_a}\right)} \tag{8}$$

where T is the recorded travel time and L the separation distance between boreholes (Kowalsky et al. 2004).

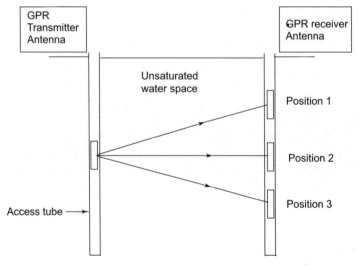

Fig. 3 Schematic GPR antenna configuration using the multi-offset gathers in the field investigation. The angle of transmitter and receiver antenna to the horizontal is limited to 45°. The position of transmitter and receiver antenna can vary.

Data is acquired by receiver in another depth borehole at some distance away. The angle of the line connecting the transmitting antenna and the receiving antenna, and the horizontal is limited to 45°. This is to avoid influences of wave reflections from high angles of antenna offset low signal to noise ratios at high angles and problems with the presence of wires within the ground. The transmitting antenna can be moved to a new position and data can be collected similarly. Repeating this process creates a dense array of intersecting ray paths, where each ray path represents the shortest path from the source to the receiver and is perpendicular to the EM wave front. First arrival time (travel time) and the amplitude of the received EM signals are used as data to evaluate the velocity and attenuation.

Weiler et al. (1998) reported that GPR, which uses an unguided EM wave, shows great promise in the future for nondestructive soil water sensing. GPR can measure larger volumes of soil than the TDR and can be utilized without disturbing the soil. This forms a clear advantage over TDR. Disadvantages include that automatic measurements are not possible because instrument that it is prone to fails in soils with high clay and salinity contents.

In GPR two modes of operations possible are the ground mode and the airborne. A strong correlation between the GPR data and the soil water content is observed in both the ground and airborne modes of operation. In the ground mode, soil moisture error is found lower than 0.03 m^3/m^3. However, in the airborne mode, soil moisture estimation is less accurate (0.046 m^3/m^3). This method has a great potential for mapping soil moisture and is efficient on most natural surface as vegetation and surface microtopography have only a small effect on the reflection of low-frequency pulses. Huisman et al., (2001) carried out GPR, TDR and gravimetric soil water content measurements. The results showed that the calibration equations between GPR and aggregated gravimetrical soil water content are similar to those obtained for TDR, suggesting that available TDR calibrations can be used for GPR.

Conclusion

The TDR method for soil water content measurement is widely applicable and is mostly used for automated data collection. However, obtaining precision and accuracy is very much dependent on wave form interpretation methods used in software (Evett, 2000). Most of these are based on the relationship between volumetric soil water content and dielectric constant (permittivity) of soils. However, a single equation is not adequate for all soils. Dirksen and Dasberg (1993) reported that this equation can be valid for the soils with low clay contents (specific surface) and typical bulk densities (1.35-1.5 g cm^{-3}). Zegelin et al. (1992) concluded that the use of universal equation gives water balance to within ±10 % of soil water content.

Each soil water sensing method has its strengths and weaknesses. A strength in one application may be a weakness in another. All of the methods have their own specific field of application. However, they complement each other in some aspects such as sensitivity at low water content. To select the right method, the user must have a good understanding of how its qualities fit the requirements of the object. Whichever sensor, may be its calibration is required to find out their site-specific behavior and plant compatibility.

References

1. Bandyopadhyay, S. (1995). Soil moisture and wheat growth assessment by microwave and optical remote sensing. Ph.D. Thesis, P.G. School, IARI, New Delhi pp. 125.

2. Bilskie, J. (1997). Using dielectric properties to measure soil water content. *Sensor Magazine,* **14:** 26–32.

3. Binley A., Winship, P., Middleton, R., Pokar, M. and West. J. (2001). High-resolution characterization of vadose zone dynamics using cross-borehole radar. *Water Res. Research,* **37:** 2639–2652.

4. Dalton F.N. and Van Genuchten, M. Th. (1986). Time domain reflectometry method for measuring soil water content and salinity. *Geoderma,* **38:** 237–250.

5. Dalton F.N., Herkelrath, W.N., Rawlins, D.S. and Rhoades. J.D. (1984). Time domain reflectometry: Simultaneous measurement of soil water content and electrical conductivity with a single probe. *Science (Washington DC)* **224:** 989–990.

6. Dalton, F.N. (1992). Development of time domain reflectometry in measuring soil water content and bulk soil electrical conductivity, pp. 143-167 in G.C. Topp, et al., Eds. Advances in measurement of soil physical properties: Bringing theory into practice. *Soil Sci. Soc. Am., Spec. Pub* l. **30**.

7. Daniels, D.J. (1996). Surface penetr;ating radar. London: Institute of Electrical Engineers.

8. Davis, J.L. and Annan, A.P. (1977). Electromagnetic detection of soil moisture: Progress report I. *Can. J. Remote Sens.* **3:** 67–73.

9. Davis, J.L. and Annan, A.P. (1989). Ground penetrating radar for high resolution mapping of soil and rock stratigraphy. *Geophys Prospect* **37:** 531–551.

10. De Loor, G.P. (1968). Dielectric properties of heterogeneous mixtures containing water. *J. Microwave Power* **3:** 67–73.

11. Dirksen C. and Dasberg, S. (1993). Improved calibration of time domain reflectomtry soil water content measurements. *Soil Sci. Soc. Am. J.,* **57:** 660–667.

12. Dirksen, C. and Hilhorst, M.A.(1995). New 20 MHz dielectric sensor for soil water content and electrical conductivity. *'College on Soil Physics' SMR.*873–6.

13. Evett, S.R. (2000). Some aspects of time domain reflectometry, neutron scattering and capacitance methods for soil water content measurement. IAEA-TECDOC-1137, 5–51.

14. Ferre, P.A., Rudilph, D.L. and Kachanoski, R.G. (1996). Spatial averaging of water content by time domain reflectometry: Implications for twin rod probes with and without dielectric coatings. *Water Resour. Res.* **32:** 271–279.

15. Hoekstra, P. and Delaney, A. (1974). Dielectric properties of soils at UHF and microwave frequencies. *J. Geophys. Res.* **79:** 1699–1708.

16. Hook W.R., Livingston, N.J., Sun, Z.R. and Hook, P.B. (1992). Remote diode shorting improves measurement of soil water by time domain reflectometry. *Soil Sci. Soc. Am. J.* **56:** 1384–1391.

17. Huisman, J.A., Sperl, C., Bouten, W. and Verstraten, J.M. (2001). Soil water content measurements at different scales: accuracy of time domain reflectometry and ground-penetrating radar. *Journal of Hydrology,* **245:** 48–58.

18. Knight, R. (2001). Ground penetrating radar for environmental applications. *Annual Review of Earth and Planetary Sci.*, **29:** 229–255.

19. Kowalsky, M.B., Finsterle, S. and Rubin, Y. (2004). Estimating flow parameter distributions using ground-penetrating radar and hydrological measurements during transient flow in the vadose zone. *Advances in Water Resources* **27:** 583–599.

20. Marshall, T.J. and Holmes, J.W. (1988). Soil Physics, Cambridge University Press, Cambridge, New York, New Rochelle, Melbourne, Sydney, p. 64.

21. Or, D. and Wraith, J.M. (2000). In Handbook of Soil Science. Editor-in- chief M.E. Sumner. pp. A53–A63. CRC Press, New York.

22. Pepin S., Livingston, N.J. and Hook, W.R. (1995). Temperature-Dependent Measurement errors in time domain reflectometry determinations of soil water. *Soil Sci. Soc. Am. J.* **59:** (1), 38–43.

23. Roth, K.R., Schulin, R., Fluhler, H., Attinger, W. (1990). Calibration of time domain reflectometry for water content measurement using a composite dielectric approach. *Water Resour Res* **26:** 2267–2273.

24. Sundara Sarma, K.S., Burman, D., Sharma, R.K. and Das, D.K. (1992). Soil moisture estimation by time domain reflectometry in saline and alkali soils. Proc. 57th Annual Convention of ISSS. Nat. Sem. Development in Soil Science, *CRIDA, Hyderabad, Nov.,* **26-29:** 3–5.

25. Topp G.C., Davis, J.L. and Annan, A.P. (1980). Electromagnetic determination of soil water content: Measurement in coaxial transmission lines. *Water Resour. Res.,* **16:** 574–582.

26. Topp, G.C. Zegelin, S.J. and White, I (1994). Monitoring soil water content using TDR. An overview of progress. In: K.M. O' Connor et al. Eds. Symposium on time domain reflectometry in environmental applications. North Western University, Evanston, IL Spec. Publ. 67–89.

27. Topp, G.C., Watt, M. and Hayhoe, H.N. (1996). Point specific measurement and monitoring of soil water content with an emphasis on TDR. *Can. J. Soil Sci.* **76:** 307–316.

28. Van Dam, R.L. and Schlager, W. (2000). Identifying causes of ground-penetrating radar reflections using time-domain reflectometry and sedimentological analyses. *Sedimentology,* **47:** 435–449.

29. Veldkamp, E. and O'Brien, J.J. (2000). Calibration of a frequency domain reflectometry sensor for humid tropical soils of volcanic origin. *Soil Sci. Soc. Am. J.,* **64:** 1549–1553.

30. Vellidis, G., Smith, M.C., Thomas, D.L. and Asmussen, L.E. (1990). Detecting wetting front movement in a sandy soil with ground penetrating radar. *Trans. ASAE* **33:** 1867–1874.

31. Weiler, K. W., Steenhuis, B.T.S., Boll, J. and Samuel Kung, K.J. (1998). Comparison of ground penetrating radar and time domain reflectometry as soil water sensors. *Soil Sci. Soc. Am. J.* **62:** 1237–1239.

32. Zegelin, S.J., White, I. and Russell, G.F. (1992). A critique of time domain reflectometry technique for determining soil water content. In: G.C. Topp et al., Eds. Advances in measurement of soil physical properties. Bringing theory into practice. *Soil Sci. Soc. Am., Madison, Wl. Spec. Pub* l. **30:** 187–208.

Topics in Electromagnetic Waves: Devices, Effects and Applications
Edited by J. Behari

17. Radiation Doses from Cosmic Rays in Aircraft Flights, Low Earth Orbit Satellites and Space Missions: A Review

A. Nagaratnam and S.C. Jain*

74A, Sita Homes, New Gyatri Nagar, Hyderabad-500 079, India

*Center for Fire, Environment & Explosive Safety,

Brig. S.K. Mazumdar Marg, Delhi-110 054, India

Abstract: The radiation doses from cosmic rays which aircrew (and frequent fliers) receive could exceed the prescribed dose limits for a member of the public, and even higher than those which nuclear reactor workers get. Of particular concern is the hazard to the fetus of a flight attendant who has got pregnant. The doses depend mainly on altitude, and also on latitude and the phase of 11-years solar cycle. Doses to astronauts and cosmonauts would be much higher, and may even reach hazardous levels. The present article reviews the components of cosmic rays, the doses likely to be received in subsonic and supersonic flights, as well as in earth satellites and space missions under various scenarios. Experience of different airlines and international regulations on the subject are indicated. Results of epidemiological surveys on the health of aircrew have been inconclusive.

Introduction

Recently we have come to realize that crews in high altitude flight of jet aircraft can receive substantial radiation doses from background cosmic radiation (which could be higher than those received even by nuclear reactor workers). As early as 1966, the International Commission on Radiological Protection (ICRP) formed a Task Group to study the radiobiological aspects of supersonic transport, which suggested that the aircrew can receive radiation doses of 1-2 mR/hr if caught in solar flares. It has also recommended that the pilots should be advised to descend to lower heights during solar flare especially in polar regions. ICRP [1] has recommended that aircraft crew should be considered as workers occupationally exposed to radiation. In addition, members of the public who frequently fly (such as couriers) may also receive exposures that are above the limits for the general public. Commercial subsonic aircraft fly at 7-12 km (23,000-40,000 ft) and supersonic aircraft at around 18 km (60,000 ft). Approximately, the dose doubles for every 1.5 km increase in altitude. Astronauts are exposed to these radiations that are much higher than for high altitude aircrews, reaching hazardous levels in certain cases. Several national and international agencies have been showing interest in such studies. This article presents the review on the subject matter and has drawn much information from the proceedings of the meeting on "Cosmic radiation exposure of aircrew, passengers and astronauts"[2].

Cosmic Rays

Cosmic Ray Environment

Cosmic rays have been studied ever since the early years of the 20th century. Cosmic radiation has three components: (i) trapped radiation, (ii) galactic cosmic radiation (GCR) and (iii) solar particle radiation (SPR).

Trapped radiations occur in two belts (called van Allen belts), one in the region of 1.2 to 2.8 earth radii, and the other in the region of 3-11 earth radii, consisting of very high energy protons and electrons. The outer zone particles are responsible for the phenomenon of aurora borealis. Practically no radiations from the trapped belts reach the earth. But they are of importance for manned missions.

Galactic cosmic radiation (GCR) is created outside the solar system and can be considered almost isotropic, and the flux is fairly constant with time. Energies cover a very wide range (10^8 to 10^{20} eV). The relative number of these particles decreases sharply with increasing energy. GCR are 98% atomic nuclei (85% protons, 12% alpha particles, and 1% heavier ions, HZE, denoting particles of high atomic number Z and high energy E) and 2% electrons. There are also a small number of photons in cosmic rays.

The GCR is influenced significantly by events happening in the sun (which itself contributes a certain amount of the cosmic radiation reaching the earth), and is also affected by the geomagnetic field, which prevents some particles from reaching the atmosphere (most effective at the equator and progressively less effective as we move towards the poles). GCR is the most significant for exposure over earth and in aircraft.

Solar Particulate Radiations (SPR), mainly high energy protons, come from the sun. Particles of very low energy are emitted continuously, but more energetic particles are emitted copiously during solar magnetic disturbances. The sun has a varying magnetic field which reverses direction every 11 years (solar cycle). The "solar minimum" occurs at the time of the reversal, when there are few sunspots, the magnetic field and solar particle emission are weak. The solar maximum causes a GCR minimum, and vice versa. Sunspots can persist for several weeks. At "solar maximum" sunspots are much more numerous, the magnetic field is large, and a plasma of electrons and protons is ejected from the sun (solar wind), which can reach the earth in half an hour. Intense bursts of high energy plasma called solar particle events (SPE) are associated with solar flares, and occur occasionally during the active period of the solar cycle. Solar flares are regions of exceptional brightness, developing suddenly, reaching maximum intensity in a few minutes, and decaying over a few hours. Generally one major event occurs every year. The largest ground level even yet observed occurred on 23 February 1956. SPE could deliver lethal doses in a few hours to astronauts.

The earth's magnetic field has a larger effect than the sun's magnetic field on the GCR approaching the atmosphere. Radiation levels are higher in polar regions and start decreasing towards the equator. The magnitude of this effect depends on altitude and time in the solar cycle.

The earth's atmosphere has a shield of roughly 10 m water equivalent at sea level (1000 g/m^2). The atmospheric depth (expressed as g/cm^2) is 1013 at sea level, 300-190 at 30,000 to 40,000 ft and 58 at 65,000 ft.

At altitudes around 10 km (those at which jet aircraft operate) and at polar latitudes, the ratio of GCR reaching the top of the atmosphere at solar minimum to that at solar maximum is 1.5-2, and increases with altitude. Again, at around 10 km altitude, the GCR flux is 2.5-5 times more intense in polar regions than near the equator. The higher the altitude, the greater is the latitude dependence. From around 50° latitude to the poles, there is no significant change in the GCR flux with latitude (polar plateau)[3].

Interactions of Primary Cosmic Rays with Constituents of the Atmosphere

When the high energy charged particles enter the top of the atmosphere, they undergo a variety of nuclear reactions with the constituents of the atmosphere, mainly carbon, nitrogen, and argon. The nucleus then breaks up into smaller units {mainly neutrons, protons, and pi mesons (pions)}. The neutrons and protons which are so formed themselves have sufficient energy to interact further with atomic nuclei and generate more particles, chiefly pions.

The most important secondary particles are positive, negative and neutral pions. The neutral pion decays into two high energy gamma ray photons, while the positive and negative pions decay into muons of the same sign. Muons can penetrate to ground level before decaying into electrons.

The photons resulting from the decay of a neutral pion can lead to creations of an electron-positron pair in the neighborhood of a nucleus. The high energy electron-positron pair, in passing through matter, loses energy through bremsstrahlung, where part of the energy of the particle is converted into photons, which, in turn, can generate electron-positron pairs, which, in turn, produce more bremsstrahlung, and so on. Thus, a neutron can initiate a cascade of secondary reactions, leading to the formation of a very large number (shower) of electrons, positrons, and photons.

The soft component of cosmic rays (which would be absorbed by 10 cm of lead) consists of electrons, positrons and photons formed from neutral pions at high altitudes and muons at lower altitudes, and is 20% of total at 50° latitude, the proportion increases at first with increasing altitude. At high altitude, the hard component consists of protons, neutrons pions and muons. At lower altitudes, the hard components consists of protons, neutrons, pions and muons. At lower altitudes, the muons are the main constituent, contributing 80% of the total dose.

Ionising Radiations: Their Sources and Levels, and Biological Effects

Ionising radiations cannot be felt by any of our senses. They require special instrumentation to be detected and measured. Ionising radiations have sufficient energy to ionize atoms and molecules, that is, pull out electrons from the field of force of the nucleus of the atom and are hazardous to biological systems. Primary cosmic rays (both particulate and photons) and their secondaries come under the category.

In addition to the inevitable natural background radiation to which we are all subjected to, over the last hundred years or so, man-made radiations have been adding to this load with increasing exposure to them.

Interaction of Radiation with Matter

Directly ionizing radiations are charged particles like alpha or beta rays. A charged particle passing through matter loses energy continuously as a result of electrostatic interactions with the electrons surrounding the atoms and molecules of the medium. The interaction may be sufficiently strong to remove on orbital electron from the atom, resulting in the formation of an ion pair, consisting of a free electron and the residual positively charged nucleus. Excitation of electrons to higher orbits around the atom also takes place. As the charged particle travels through the medium, it continuously loses energy by these processes. The specific ionization is the number of ion pairs produced per unit length of the tract of the ionizing particle (in air, around 50,000 per cm for alpha particles and 30-300 for betas). The linear energy transfer, LET (expressed as keV per cm of air or micron in water) is the rate of energy transfer from the ionizing particle to the medium.

Since indirectly ionizing radiations like X-rays, gamma rays or neutrons do not carry any charge, they cannot produce ionization by direct interactions with matter, but produce secondary charged particles (electrons in the case of X- and gamma-rays, and mainly recoil protons in the case of neutrons) which are directly ionizing.

Biological Effects of Radiation

The first step in the interaction of radiation with matter (whether inorganic or organic) is the deposition of a certain amount of energy from the radiation to the target in an extremely short time. This leads to the excitation and ionization in the atoms and molecules exposed, followed by formation of highly reactive short-lived free radicals, some of which diffuse out of the site of formation to interact with sensitive constituents of the cell, in particular, DNA, inside the nucleus. In addition to this indirect damage, DNA can also suffer direct damage as a result of a direct hit with ionizing radiation (single strand and double strand breaks, chromosomal aberrations). The body has several powerful enzymatic mechanisms to repair the damage most of the time.

When a cell is damaged, one of three things can happen. The repair may be complete and faithful, in which case nothing happens to the organism. Or, there may be non-repair, in which case the affected cell dies after a few divisions. Since our organs each contain of the order of 10^{11} cells, the body can withstand destruction of a fraction of the cells without suffering undue results unless a substantial fraction of the cells is killed (from higher doses of radiation), manifesting in the loss of function or even death. The third alternative is misrepair, where the cell is not killed, is viable, but has altered genes. This is the most serious type of damage, because the viable or misrepaired cell would continue to divide and produce wrong copies of the cell; this may lead to the induction of cancers after a long latent period (years to decades), or if the germinal (gonadal) cells are affected, lead to the transmission of the wrong cell to the offspring during fertilization, resulting in potential hereditary or genetic effects.

Stochastic and Deterministic Effect

Deterministic effects are those for which the severity of the effect increases with the dose and for which there is normally a well-defined threshold (a few Gy). Examples are cataract of the lens of the eye, disorders of the blood system like aplastic anaemia and impairment of fertility (temporary or permanent). So long as the threshold dose is not approached, the person will not exhibit the particular deterministic effect. Deterministic effects are due to cell killing, the severity of the effect depending on the proportion of cells of the organ or tissue killed.

Stochastic effects are of a probabilistic or random nature. The main types of such effects of concern are the induction of leukemia and other cancers, as well as hereditary effects. These arise as a result of a single cell or a small number of cells being damaged, leading to viable but misrepaired cells which proliferate with the wrong copy being passed on from generation to generation of cells during their division. It is postulated that there is no threshold to the induction of stochastic effects. That means that, unlike in the case of deterministic effects, there is no safe dose below which the effect would not occur. Even the lowest dose has its own small probability of producing the effect. Further, the severity of the effect does not depend on the dose. Thus, whether leukacmia is induced from exposure to 1 Gy or 3 Gy, the disease will run its own fatal course. The higher the dose, the greater is the probability of the effect occurring.

Acute radiation syndrome: An important class of deterministic effect is the acute radiation sickness from acute whole body exposure (received over a very short period of time, say minutes), as happened in the case of the Japanese atomic bomb victims. Depending on the dose, the victims exhibited symptoms (gastrointestinal and blood disorders). At a dose of 2 Gy, the main symptom will be nausea, but hardly 5% of those exposed will die. AT 3-5 Gy, nausea and vomitting occur within a few hours, followed by a latent period of well-being for about a week, followed by serious complications of the blood system; death would occur in 50% of exposed individuals within 30 days at a dose of 5 Gy. At higher radiation doses, practically everyone exposed would die within a week.

Stochastic effects: The most important class of stochastic effects is the induction of cancers. Table 1 gives the latest estimates of cancer induction from a chronic whole body exposure of 1 Gy[1, 4]. Another important case of concern relates to in-utero exposures; there is higher probability of childhood (within 10 years of birth) leukaemia and other cancers in children born to mothers who underwent pelvic diagnostic X-ray examinations (doses as low as mGy) during their pregnancy.

Genetic or hereditary effects: When the germ cells in the gonads (testes and ovaries) are mutated as a result of exposure to radiation, such a mutated cell, when it takes part in fertilization, will lead to a zygote (formed by the union of maternal and paternal germ cells) carrying the mutated cell. The offspring would carry the mutated cell, whose effects can vary very widely, from practically no effect on the health and longevity of the offspring. Mutations occur naturally also; they are, by and large, harmful.

We do not have any direct evidence on the genetic effects of radiation in the human. However, based on evidence from animal studies, we make the cautious, conservative assumption that such hereditary effects can also occur in humans. Table 1 also gives the risk of serious genetic effects for a chronic exposure of 1 Gy.

Table 1.   Risk coefficients for radiation effects [1, 4] chronic exposure to 1 Sv; population of all ages

Lifetime mortality risk	
Solid cancers:	
Male	4.5%
Female	6.5%
Incidence risk double the mortality risk	
Leukaemia: Male and Female 0.5%	
Incidence and mortality risks identical	
Risk for children twice that for adults	
Risk for acute exposure: Twice that for chronic exposure	
Uncertainty factor: ~ 2	
Health detriment from non-fatal cancers: 1.0%	
Severe hereditary defects: 1.3%	

Several factors affect the radiation response of biological systems, among which are total dose, dose rate, nature of the radiation, relative radio-sensitivity of organs and tissues, and other factors. The lower the dose rate, the lower the biological effect. (This is because at low dose rates, the body has enough time to repair the damage.)

Protection from Excessive Exposure to Radiations

Mankind has been reaping invaluable benefits from the peaceful uses of nuclear energy. On the other hand, excessive exposure to radiation can cause harmful biological effects. In this respect radiation is no different from other risks in modern life, since all facets of human advance almost invariably entail some concomitant risks. Wisdom lies in striking a happy and judicious balance in order to ensure that we continue to utilize radiation for human benefit while keeping the potential risks to an acceptable minimum. ICRP states that the procedures available to control exposures are sufficient, if used properly to ensure that radiation remains a minor component of the spectrum of risks to which we all are exposed.

ICRP has, after a great deal of thought, laid down separate dose limits for occupational exposure (20 mSv/year effective dose) and public exposure (1 mSv/year).

Natural Background and Man-made Radiation

Natural background radiation consists of three major parts, viz. cosmic rays, terrestrial radiation and body radioactivity.

Cosmic rays have already been discussed in detail.

Terrestrial radiation: The earth's crust contains some important radionuclides, the most important of which are the uranium and thorium series; and radioactive potassium, ^{40}K, which occurs in natural potassium to the extent of 0.018%. Building materials, which are made from soil and rocks, would also contain these radionuclides. All these emit gamma rays, and therefore we can receive a radiation dose from soil, rock and building materials. One of the daughter products of the uranium and thorium series is radon, a gas, which is present in environmental air (both indoors and outdoors).

The body contains several radionuclides in very minute amounts, chiefly products of the uranium and thorium series and ^{40}K.

Doses from Natural Background Radiation [4]

Table 2 summarizes the average worldwide per capita annual dose from natural background radiation, as well as the normal range of variations [4]. This average

Table 2.   Radiation doses from natural sources (world average) [4]

Source	Worldwide average per capita annual effective dose (mSv)	Typical range (mSv)
External exposure		
Cosmic rays	0.4 (17%)	0.3-1.0[a]
Terrestrial gamma rays	0.5 (21%)	0.3-0.6[b]
Internal exposure		
Inhalation (mainly radon)	1.2 (50%)	0.2-10[c]
Ingestion	0.3 (12%)	0.2-0.8[d]
Total	2.4 (100%)	1-10

[a] Range from sea level to high ground elevation.

[b] Depending on radionuclide composition of soil and building materials.

[c] Depending on indoor accumulation of radon gas.

[d] Depending on radionuclide composition of foods and drinking water.

dose comes about to 2.4 mSv, with the actual values likely to range from 1 to 10 mSv. Inhalation of randon and its daughter products contributes the largest fraction (50%) to the total; next comes terrestrial radiation (21%) followed by cosmic rays (17%), and ingestion (12%).

Man-made radiations: Man-made radiations may be broadly classified as follows:

(i) ***Medical:*** This is the most important category comprising diagnostic radiology, radiotherapy and nuclear medicine.

(ii) ***Occupational exposure:*** Like workers in uranium mines or nuclear fuel cycle operations, those concerned with medical or industrial applications of radiation, defence activities (chiefly production and maintenance of nuclear weapons and nuclear submarines, accidents and fallout from nuclear weapons tests.

Another important category of occupational workers (this includes aircraft crew) are those exposed to so-called enhanced natural radiation sources. It refers to the natural background radiation levels in certain situations and locations, where human activities may lead to appreciable higher levels of exposure than normally encountered. These primarily include: Locations with much higher levels of radon, and Air and Space travel.

The average doses to those exposed occupationally to man-made radiations and enhanced natural radiations are 0.6 and 1.8 mSv, respectively. The average exposure to 250,000 air crew is 3.0 mSv, as against 1.8 mSv for nuclear cycle workers [4].

The total worldwide average per capita effective dose in the year 2000 from both natural and man-made radiant sources is 2.8 mSv: 86% from natural radiations, 14% from diagnostic medical examinations, 0.2% from atmospheric nuclear testing, < 0.1% from the Chernobyl accident, and < 0.01% from nuclear power production (at the current rate of 250 GW-y) [4].

Doses from Cosmic Rays

Both experimental and theoretical techniques are used for estimation of cosmic ray doses. Several codes have been developed for dose computations, one of the most widely used being CARI-6, developed by the US Federal Aviation Administration, which can estimate for a given flight profile the effective dose received on a non-stop flight between any two locations in the world at altitudes up to 87,000 ft. On-line information on solar activity is available from a variety of websites [5].

The mean rate of ion production from cosmic rays at sea level is $2.1 \text{ cm}^2 \text{ s}^{-1}$ corresponding to a dose rate of 32 nGy.h^{-1} (or 32 nSv.h^{-1}, with a radiation weighting factor of 1). With a mean shielding factor of 0.8 indoor dose rate comes to 26 nSv.h^{-1}. Assuming average fraction of time spent indoor as 0.8, annual effective dose rate from ionizing component at sea level is 240 μSv. Taking into account the relative populations living at different latitudes and altitudes, the population-weighted average dose from the directly ionizing and photon components for the world comes out to 280 μSv. The contribution from the neutron component is 100 μSv. The total population-weighted annual dose thus comes out to 380 μSv (range 300-3000 μSv) for the majority of the world population.

At ground level, muon component contributes 80% of dose from directly ionizing radiation; the remainder comes from electrons. At aircraft altitudes n, e^+, e^-, p are important. At still higher altitudes, heavy nuclei would have to be considered.

The exposure of crew members is expressed as the time between leaving the terminal before takeoff and returning after landing. Thus, the block hours denote "gate-to-gate" time and not flight time. For flights of more than one hour, exposure rate at cruising altitude will be the main determinant of the dose. Annual number of flying hours flown by crew members of subsonic aircraft is in the range of 300-900 hours per year (average 500–600). Supersonic aircraft crew fly about half this time. For the general population, 3 groups can be identified: non-flyers (0 h), occasional fliers (10 h average, 3-50 h range), and frequent fliers (100 h average, 50-1200 h range). Doses would be higher in polar flights as compared to equatorial flights.

Crew exposures are fairly protracted and at low dose rate. The main contributor to the dose is galactic cosmic radiation (GCR) and its secondaries. Solar particle events would normally make only a minor contribution for subsonic flights. However, a worst case SPE (like that of 23-3-1956) can give several mSv effective dose for a long distance high latitude subsonic flight, and greater than 100 mSv for a high latitude supersonic flight. The pilot will have to descent to a lower altitude when the monitor alarm indicates a high dose rate.

The mean effective dose rates for an equatorial flight increases from 0.5 $\mu Sv.h^{-1}$ at 20,000 ft to 3.2 $\mu Sv.h^{-1}$ at 80,000 ft; the corresponding values for a high latitude flight are 1.1 $\mu Sv.h^{-1}$ and 17 $\mu Sv.h^{-1}$ respectively [6].

Most estimates by different authors agree that the cosmic ray exposure of most present jet aircraft crew exceeds 1 $mSv.y^{-1}$ (which is the ICRP limit for a member of the public).

Figure 1 gives dose rates as a function of various component particles of GCR near the polar plateau and solar minimum [7].

The average exposure of a commercial pilot on a polar flight in a subsonic aircraft is around 100 μSv per flight. Taking a mean flight time of 600-700 block hours per year at 10 km and above latitudes, estimated effective doses to subsonic aircraft crew would be 5–9 $mSv.y^{-1}$. For a pregnant crew member, for a work schedule of 70 block hours a month, the monthly dose would be around 0.45 mSv; in two months the prescribed dose limit of 1 mSv to the fetus would be reached [6].

A passenger traveling for 200 hours in a subsonic aircraft is likely to receive an annual effective dose of 1 mSv and is hence not of special concern. Effective doses for supersonic aircraft (10-12 $\mu Sv.h^{-1}$) would be double the values for subsonic aircraft, and the annual exposures may exceed the occupational exposure limit of 20 $\mu Sv.y^{-1}$.

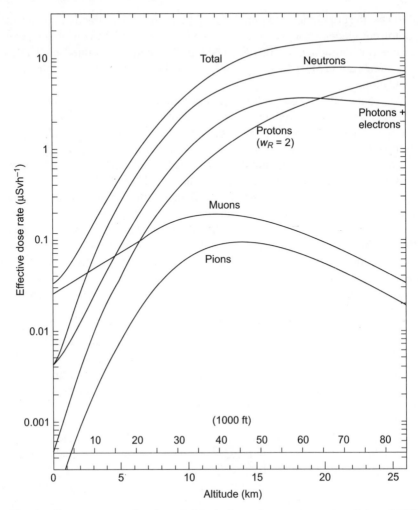

Fig. 1   Dose rates as a function of altitude for various component particles of GCR near the polar plateau and solar minimum (Reproduced from the journal *Health Physics* with permission from the Health Physics Society).

Experiences of Different Airlines

A total of around 1.8×10^{-12} passenger-hours were flown throughout the world (3×10^{9} passenger-hours aloft). US airlines have 167,000 flight personnel (including 96,000 flight attendants).

The experiences of airlines from different countries are broadly the same, and are discussed in several papers in the November, 2000 issue of Health Physics [7-12].

The US Federal Aviation Administration has provided instruction manuals that contain information on estimates of doses for a variety of flights, and tables for estimating health risks from various doses, including doses to pregnant women.

The British Concorde that entered into service in 1976, flew at ~18 km altitude. Monitoring equipment (with several types of detectors) were installed in all Concords along with an audio alarm and a light is the master warning system. Passengers got 1 mSv per year for 100 h of flight in Concords and 200 in transequatorial routes [2].

Average doses to flight crew, cabin attendants and duty travellers do not exceed 6 mSv.y^{-1}. Frequent travellers may exceed 1 mSv.y^{-1} if flying more than 6-8 times per year.

Table 3 gives estimates [4] of effective doses from cosmic radiation for typical flight routes, as reported by different concerned agencies. The values vary from 0.2 to 9 mSv.y^{-1} for annual doses to aircrew [2].

Current Regulations and Recommendations

Different airways have slightly different regulations. It is generally accepted that monitoring is not required for individuals whose expected annual exposures are less than 1 mSv. On-board monitoring may be required where doses may exceed 6 mSv.y^{-1}.

The International Civil Aviation Organisation, ICAO, in 1995 recommended that 'all airplanes intended to be operated above 15,000 m (49,000 ft) shall carry equipment to measure and indicate continuously the dose rate of total cosmic radiation received (total of ionizing and neutron radiation from both GCR and SPE) and the cumulative dose on each flight. The display unit of the equipment shall be readily visible to a flight crew member".

Special recommendations have been made by ICAO and ICRP regarding pregnant members of the aircrew. Many airlines take flight attendants off duty for the period of pregnancy.

Astronauts and Space Missions [13]

Quite a number of measurements have been made in US Shuttle Space Missions, in the MIR Orbital Station, and missions to Mars.

Low Earth Orbits (LEO)

The exposures incurred in low earth orbit (LEO) are from trapped radiation, GCR and SPE. The expected doses in LEOs of low and high inclinations are 20 and 150 μSv, respectively.

In US Space Shuttle missions, the average astronaut doses were in the range of 100 μSv per day (range 30-220 μSv.day^{-1} depending on mission).

Deep Space Missions

Outside the geomagnetic sphere astronauts are subjected to full GCR and SPE, although the dominant contribution is orally from GCR. The average doses are expected to be 500 μSv per mission.

Table 3. Estimates of effective doses from cosmic radiation for typical flight routes [4]

Route	Flight duration (min)	Effective dose (mSv)	
		One flight on route	1000 h on route
Short-haul routes			
Dublin-Paris	95	0.0045	2.8
London-Rome	135	0.0067	3.0
Frankfurt-Helsinki	160	0.0100	3.7
Brussels-Athens	195	0.0098	3.0
Luxembourg-Madrid	130	0.0054	2.6
Stockholm-Vienna	140	0.0082	3.5
Lisbon-Munich	180	0.0091	3.0
Copenhagen-Dublin	120	0.0071	3.5
Amsterdam-Manchester	70	0.0030	2.6
Dublin-Rome	180	0.010	3.3
Long-haul routes			
Stockholm-Tokyo	605	0.051	5.0
Dublin-New York	450	0.046	6.1
Paris-Rio de Janeiro	675	0.026	2.3
Frankfurt-Bangkok	630	0.030	2.9
London-Toronto	490	0.050	6.2
Amsterdam-Vancouver	645	0.070	6.6
Los Angeles-Auckland	760	0.030	2.3
London-Johannesburg	655	0.025	2.3
Perth-Harare	665	0.039	3.5
Brussels-Singapore	675	0.030	2.7

Total doses for a Mars mission of 2.5 years (with 6 months transit time each way and 1.5 years' stay in Mars) have been computed to be 500 mSv at solar minimum and 1200 mSv at solar maximum.

Health Effects and Epidemiological Studies on Aircrew

Epidemiological studies have been conducted over the last 20 years on pilots and cabin crew (World War I and II military pilots, Korean conflict pilots, and commercial pilots). A recent major European study has involved 22,500 cockpit crew and 50,000 civilian crew from 9 countries [14].

No Acute effects have been documented so far in pilots, flight attendants, or cosmonauts. The results of the various surveys are not always concordant. The conclusion is that for delayed effect (like cancer) no clear picture has emerged with regard to disease patterns (except a lower risk of total deaths in comparison with the general population, and a high risk of death from airplane accidents).

Conclusion

Sufficient theoretical protocols and experimental techniques are available to make fairly reliable estimates of doses from cosmic rays (which include trapped radiation, galactic cosmic radiation and solar particle events) to air crew in subsonic and supersonic flights. The doses increase rapidly with altitude; they are somewhat higher at polar latitudes than equatorial ones; they also depend on the solar activity at the time of flight. A careful decision has to be made whether a particular member of the air crew needs to be considered as a worker occupationally exposed to radiation. A flight attendant who has become pregnant is a special case that needs vigilance.

The doses could be much higher for astronauts and cosmonauts, reaching hazardous levels under certain atmospheric conditions. National agencies of many countries, as well as the International Civil Aviation Organization, have formulated safety protocols. Epidemiological surveys on health of air crew have not demonstrated any adverse effects.

So far as India is concerned, both Indian Airlines and Air India have only subsonic aircraft; the former fly mainly along equatorial routes, while the latter fly along both equatorial and polar routes. The situation is therefore fairly satisfactory. Indian Air Force has several squadrons of supersonic aircraft (which fly mainly in the equatorial regions). It may be worthwhile to examine the doses likely to be received by Indian Air Force pilots.

References

1. I.C.R.P. Recommendations of the International Commission on Radiological Protection. ICRP Publication 60, *Annals of the ICRP*, **21** (1–3), Pergamon Press, Oxford, 1991.

2. Proceedings of 1998 meeting on "Cosmic radiation exposure of aircrew, passengers and astronauts" Special Issue. *Health Phys.* **79**(5), 465–613, 2000.

3. Wilson, J.W. Overview of radiation environments and human exposure, *Health Phys.* **79** (5), 470–494, 2000.

4. UNSCEAR. Sources and Effects of Ionizing Radiation. Report of the United Nations Scientific Committee on the Effects of Atomic Radiation, United Nations, New York, 2000.

5. News on the Net. Solar activity. *Health Phys.*, **79**(5), 465, 2000.

6. Friedberg, W., Copeland, K. Duke, F.E., O'Brien, K. III, Darden, E.B. Jr. Radiation exposure during air travel: Guidance provided by the Federal Aviation Administration for air carrier crew. *Health Phys.*, **79**(5), 591–595, 2000.

7. Goldhagen, P. Overview of aircraft radiation exposure and recent ER-2 measurements. *Health Phys.*, **79**(5), 526–544, 2000.

8. Bagshaw, M. "British Airways measurement of cosmic radiation exposure on Concorde supersonic transport. *Health Phys.* **79**(5), 545–546, 2000.

9. Chee, P.A., Braby, L.A., and Conroy, T.J. Potential doses to passengers and crew of supersonic transports. *Health Phys.* **79**(5), 547–552, 2000.

10. Menzel. H.G., O'Sullivan, D., Beck, P. and Bartlett, D. European measurements of aircraft crew exposure to cosmic radiation. *Health Phys.*, **79**(5), 563–567, 2000.

11. Tume, P., Lewis, B.J. Bennett, L.G.I., Pierre, M., Cousins, T., Hoffarth, B.E., Jones, T.A., and Brisson, J.R. Assessment of the cosmic radiation exposures on Canadian-based routes, *Health Phys.*, **79**(5), 568–575, 2000.

12. Carter, E. Perspectives of those impacted: Flight attendant's perspective. *Health Phys.*, **79**(5), 600–601, 2000.

13. Badhwar, G.D. Radiation measurements in low earth orbits: US and Russian results. *Health Phys.*, **79**(5), 507–514, 2000.

14. Boice, J.D., Jr., Blettner, M., and Auvinen, A. Epidemiological studies of pilots and aircrew. *Health Phys.*, **79**(5), 576–584, 2000.

Topics in Electromagnetic Waves: Devices, Effects and Applications
Edited by J. Behari
Copyright © 2005, Anamaya Publishers, New Delhi, India

18. Modern Applications of Lasers

K.N. Chopra

Laser Science and Technology Centre, Metcalfe House, Delhi, India

Abstract: With the advent of lasers, its applications in all fields of science and technology have been diversed, extraordinary and unique. Lasers are extraordinary tools to modify the fundamental properties of metals and also used for identifying materials. High power lasers are used for cutting of metals and also for surface hardening, annealing, soldering and drilling of precise holes in metallic shields. They are also used for document scanning, type setting, newspaper plate making, memory devices and labeling of complex mechanical parts. In addition, lasers are used in bar code scanners for retail stores and super market checkout stands. These are also used for surveying and range finding. In addition, lasers find many applications in medicine. This article describes briefly the laser applications in various fields.

Introduction

Just as nature by periodic mutation creates new life-forms and moves the species to new levels of performance, mutative cerebration in human endeavors provides new concepts that can refine and develop into processes that enrich the human condition. The laser is such a type of contribution whose impact is already widespread in science, communication, industry and medicine. This impact will grow rapidly as we better understand its nature and integrate both basic and advanced forms closely with other core technologies to produce hybrids that can satisfy as yet unknown needs. Some of the laser applications in various fields are described as follows.

Laser Radar

Laser radar technology is an obvious progression of radar technology from the radio frequency, microwave, and millimeter wave regions of the spectrum into the optical region (i.e. infrared, visible, near ultraviolet). Laser radar systems are used predominantly in those applications that cannot be addressed by conventional radar systems. Range finders are the most basic radar systems of either the microwave or the laser variety. They measure the range to a target by measuring the time of flight of a transmitted and an echo pulse of electromagnetic radiation. The speed of the target can be obtained by measuring the change in range as a function of time. Range finders can also provide information about the azimuth to a target.

Coherent radar systems are more complex and have the ability to measure the velocity of the targets by means of the Doppler effect. Coherent radar systems measure the Doppler shift of the echo radiation by comparing the frequency of the received echo signal with the frequency of the transmitted radiation. This comparison is accomplished by heterodyning, or mixing the returned signal with the signal of the system's frequency reference (called the local oscillator) on a detector. By maintaining the frequency of the transmitter signal either above or below the local oscillator signal by a fixed value determined by electronic control circuits, and superimposing on the detector the local oscillator signal with the return signal from a stationary target, an interference radiation is generated on the detector. This "beat" signal is equal to the difference between the frequencies of the transmitted and local oscillator signals. Since this beat signal is arranged to occur within the radio frequency range (tens to hundreds of mega hertz), electronic amplifiers tuned to this frequency can process the signal electronically and obtain the same signal-to-noise benefits well known in conventional heterodyne radio receivers.

Measurement of the deviation of this known beat signal by the Doppler effect caused by the moving target provides a measurement of the speed of the target. An additional advantage of the coherent radar system is that one can increase the power of the local oscillator signal on the detector to achieve the theoretical detector performance, which is the quantum noise-limited sensitivity. A laser radar, using CO_2 laser, typically operates in the 10.6 mm wavelength region. At this wavelength, HgCdTe detectors at present provide the optimum sensitivity. A merit for detectors is usually given in terms of noise-equivalent power (NEP). Heterodyne NEPs of 2×10^{-19}, 5×10^{-18}, and 2×10^{517} WHz have been measured with HgCdTe detectors operating at 1 GHz at 77 K, 195 K, and 300 K, respectively.

Table 1 compares some of the relevant parameters of x-band and CO_2 laser radars. The laser radar operates at a frequency 3,000 times higher (or at a wavelength 3,000 times shorter) than an x-band radar. The large difference in wavelengths between the CO_2 laser radar and the x-band radar results in large differences in reflection characteristics of targets for the two technologies. Variation in target surface dimensions (i.e. surface roughness) typically are greater than the wavelengths of CO_2 laser radars, but less than the wavelengths of x-band radars. Since man-made targets usually have smoother surfaces than natural targets, even small man-made targets such as wires have a larger cross-section than natural targets for laser radars. Since the beam divergence varies directly with wavelength and indirectly with transmitting aperture, CO_2 laser radars have 3,000 times smaller beam divergence than x-band radars with the same aperture.

Table 1. Comparison of basic radar parameters

Radar Characteristics	CO_2 Laser Radar	x-Band Radar
Frequency, Hz	3×10^{13}	10^{10}
Wavelength, cm	10^{-3}	3
Beamwidth, λ/D, radians	10^{-3}/dia	3/dia
Doppler sensitivity, $2v/\lambda$,Hz	$2,000 \times$ velocity	$2/3 \times$ velocity
Photon energy, Joules	2×10^{-20}	6.6×10^{-24}

Communications

The explosion in the use of lasers in communications has come about through a simultaneous improvement in the quality of the medium through which light energy is transmitted and the increased understanding of the laser sources, detectors, and associated phenomena that allow tailoring of the properties of materials and devices. For economic exploitation of fiber-based lightwave systems for information transmission, two parameters, sometimes combined, are very important. The first is the maximum data transmission rate, which itself is limited by capabilities of the laser, detectors, and associated electronics. The second is the maximum distance a bit stream can be transmitted over an optical fiber before a repeater is necessary. It is clear that the properties of the optical fiber contribute to the second parameter. Relevant optical fiber properties are the absorption losses and the chromatic dispersion. An appropriate standard of measurement for an information transmission system is the product of bit rate and distance, that is, the distance between repeaters at a prescribed bit rate. The introduction of lightwave systems is causing a sharp change in the rate at which channel capacity has increased over the past 100 years.

Laser Pumped Optical Parametric Oscillator

Optical Parametric Oscillators (OPO) as shown in (Fig. 1) are increasingly being used to extend the frequency range of existing laser sources and are well suited for the development of compact Eye Safe Laser sources. Efficient conversion of existing Nd:YAG laser based range finders to eye safe region and provision of switchable eye safe range finding option in target designators are being realized with OPO technique. High conversion efficiency and low beam divergence OPO design is perquisite for these applications due to overall system weight and size restrictions. For generating fixed eye safe wavelength, OPO construction is usually based on Non Critically Phase Matched (NCPM)-KTP crystal due to its high effective non-linear coefficient and high damage threshold.

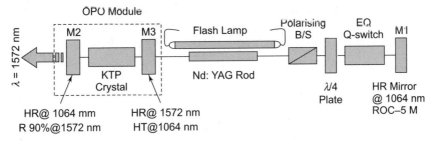

Fig. 1 Optical scheme of intracavity optical parametric oscillator.

For converting 50-100 mJ level of output energies typically generated from Nd:YAG laser based target designators, extra cavity pumped OPO (EOPO) are being used efficiently. Typical threshold power density of KTP crystal based OPO is around 11 MW/cm^2. For achieving maximum conversion efficiency,

5 to 6 times the threshold power density is pumped into the OPO. These levels of power densities for 50 to 100 m J pulse energies in 20 nsec pulse width can be realized by reducing the pump beam size to around 2.5-3 mm in the KTP crystal. For these focused spot sizes, the depth of focus is sufficiently long to allow mode matching of the plane-parallel resonator with that of the focused pump beam thereby resulting in lower OPO output beam divergences. For converting 10-15 mJ of output energies generated from hand held type Nd:YAG laser range finders, maximum conversion efficiency requirement leads to a focused pump beam size of less than 1 mm in the crystal thereby resulting in lower depth of focus. Switching over to confocal resonator becomes necessary to satisfy the mode matching requirements, resulting in increased beam divergence.

Surface Treatment

The lasers are used to harden the surface of metals through rapid heating and quenching of a surface layer. Surface mechanical properties (hardness, abrasion, resistance etc.) can often be greatly enhanced through the metallurgical reactions produced during these heating and cooling cycles. Steels and cast irons are particularly good candidates for laser transformation hardening. The process has unique advantages, particularly when used to enhance surface properties in local areas without affecting other areas of the component surface. At room temperature, plain carbon steels contain a mixture of a body-centred cubic phase (ferrite) and an iron carbide phase. Upon heating to a high enough temperature, the carbides and ferrite dissolve into a single face-centered cubic phase called austenite. The temperatures involved are usually between 750°C and 1000°C depending on the chemistry of the steel.

When a laser beam with high irradiance is rapidly scanned over a metal surface, it produces a thin layer of molten material near the surface. High energy density leads to localized surface of melting with efficient use of the energy for melting. Thus, almost all the energy is used for melting, and only a small amount is lost to subsurface heating. One maintains a cold subsurface while melting a thin surface layer. This leads to very rapid quenching of the molten material by conduction into the substrate after the end of the short irradiation.

Removal of Paint, Dielectric and Other Coatings

Laser ablation provides a method of selectively removing coatings with great precision from delicate substrates without causing damage, even when substrate and coating(s) are nearly indistinguishable from each other. Also, the ejectant residue (sometimes hazardous) may be completely and safely captured. The recent development of special rep-pulsed CO_2 lasers, which combine high peak and average powers (1) (up to 50 MW peak and 10 kW average, respectively), makes it feasible to strip very large, arbitrarily shaped objects such as airplanes, ships, buildings, bridges, and oil derricks at speeds.

Brazing/Soldering

Brazing is defined as a process where the base material remains in the solid state and a solder material reaches the liquid state, thus wetting the base material. Brazing is divided into soft brazing (temperature below 450°C) and hard brazing (temperature above 450°C). Brazing is done generally with a base metal, a braze metal, and in many cases a flux, which destroys and removes the oxides from the base metal and protects the base metal and the solder from oxidation during the process. The fluxes can cause corrosion and then have to be removed carefully from the brazed components. Fluxes and braze metals are often unhealthy for workers, and safety must be considered. The most important requirements for achieving high-quality joints are clean surfaces, normally inside a small gap, and guaranteed working temperature over the whole brazing area. Under these conditions, the solder material can flow into the gap because of capillary forces and wet the surfaces.

Conduction Welding and Penetration Welding

Laser welding is used when it is essential to limit the size of the heat-affected zone, to reduce the roughness of the welded surface and to eliminate mechanical effects. Solid-state lasers operating in the continuous or pulsed mode can function as welding sources. At higher values of laser power, a keyhole may be formed in the material and the laser energy may be deposited deeper in the workpiece. This allows production of welds having greater depth. Precisely focused laser beams are used for the welding of materials. Depending on the laser irradiance, either a conduction-limited or a deep-penetration mode is possible. Conduction-limited welds are obtained in metals at an irradiance less than 10^6 W/cm^2. At these irradiance levels, the focused laser beam is absorbed by the metal workpiece, generating heat that is rapidly conducted into the metal, thus melting the laser-irradiated surface and the subsurface layers. This process goes on without vaporization of molten surface layers.

Laser Cutting

Laser cutting, the most established laser materials processing technology, is a method for shaping and separating a workpiece into segments of desired geometry. The cutting process is executed by moving a focused laser beam along the surface of the workpiece with constant distance, thereby generating a narrow (typically some tenths of a millimeter) cut kerf. This kerf fully penetrates the material along the desired cut contour.

Balancing

Balancing is a procedure, by which the mass distribution of a rotor is checked and, if necessary, adjusted to ensure that the residual unbalance or the vibration of the journals and/or forces on the bearings at a frequency corresponding to service speed are within specified limits. Typical metal working machines apply the same CNC program for all parts of a type. Balancing machines have to treat each part (rotor) separately, since unbalances are individually created by scattered tolerances in dimensions, materials, and fits. The aim of balancing is to let the shaft axis (given by the bearings) and the mass axis (principal axis of interia near to the shaft axis) coincide within a permissible error, the balance tolerance. This may be performed either by moving the shaft axis toward the mass axis or by moving the mass axis toward the shaft axis.

Laser Marking/Branding

Laser marking or branding is used to mark the complete range of integrated circuit packages manufactured by the semiconductor industry. Major benefits are permanence of the mark, ease of programmability of the marking information, and compatibility with computer integrated manufacturing (CIM) environments. Laser-based systems are reliable manufacturing tools and provide consistent process control with low maintenance and service requirements. Lasers are also environmentally friendly and reduce or eliminate regulatory requirements over existing processes.

Laser-Based Rapid Prototyping

Rapid prototyping (RP) is a description applied to a set of technology that fabricates components through layered manufacturing (i.e. one layer at a time, succeeding layers adhered to preceding layers). RP systems form solid material only where it exists in the component. In comparison, traditional manufacturing starts with a block of material and cuts away all the excess material.

Link Cutting/Making

Laser cutting of links is accomplished by vaporization of the conducting lines which connect the circuit elements. The links are polysilicon or metals like gold, copper, and aluminum, with thickness of a few micrometers. Lasers that have been used for link cutting have included the argon ion laser, the Nd:YAG laser, and excimer lasers.

Laser-Based Photomask Repair

Photomasks used in integrated circuit production are formed by patterns of chromium metallization on clear glass substrates. Photomasks may contain a variety of different types of defects. Defects include both excess chromium metallization, either as isolated spots or as extensions of metal from a feature, and also areas with missing metallization. The ability to repair such defects provides a cost-saving alternative to discarding the defective mask. Repair leads to higher productivity and faster turn around in the mask making process.

Photolithography

Lasers are used as an illumination source for deep ultraviolet ($\lambda <$ 0.25 mm) lithography, where the i-line of a Hg lamp ($\lambda = 365$ nm) is not sufficiently short to produce semiconductor chips with critical feature sizes smaller than 0.25 μm. A typical application concerns photolithography for manufacturing 256 Mbit DRAM (dynamic random access memory) chips. Fabrication of chips with critical geometries down to 0.18-0.15 mm is carried out with band-narrowed KrF excimer laser ($\lambda = 248$ nm). The general characteristics of KrF laser used for deep ultraviolet photolithography are

Laser energy	:	~10 mJ/pulse
Reception rate	:	1000 Hz
Spectral width	:	< 0.8 pm at FWHM

Integrated circuits with geometries smaller than 0.15 μm require shorter-wavelength photolithography, such as that based on ArF ($\lambda = 157$ bn) lasers.

Laser Produced Microstructures

Laser microstructuring is a method of fabricating small feature size structures, typically 100 μm, or less, using laser beams from the near infrared (1.06 μm, Nd:YAG) to deep-ultraviolet (157 nm, F_2 excimer). Laser beams of Raman shifted radiation at wavelengths below 150 nm have also been applied for microstructuring. The interaction of short duration pulses (100 ns or less) with solids results in a relatively confined heat-affected zone. Precision micromachining with few exceptions is carried out with ns, ps, or fs pulses. Microstructuring that involves material conditioning (annealing, surface texturing, hardening, etc.) and/or laser-assisted growth is typically carried out with longer pulses or with CW lasers.

A laser writing approach makes use of a tightly focused beam (Nd:YAG, Ar^+, HeCd, excimer) that is delivered to the surface of a processed piece with a focusing lens. Beam expansion and spatial filtering are applied to achieve diffraction-limited spots. Complicated 3D shapes can be obtained by moving the workpiece on a digitally controlled x-y-z stage and/or by steering the laser beam with numerically controlled mirrors.

Electronic Packaging

Packaging in microelectronics involves the integration of chips into a complete assembly to perform a specific function. A good example is the assembly of multichip modules that are collection of integrated circuits mounted onto a high-density interconnect substrate. The multichip module can require thousands of electrical connections, which have to be produced reliably and inexpensively within a very small space. As chip geometries continue to shrink, the performance of a system becomes limited by the packaging and interconnect technology.

Biotechnology

Biotechnology is a major growth area for many electro-optic technologies and the helium neon laser is certainly no exception. HeNe lasers are now used extensively in both laboratory research as well as in analytical instruments for important clinical tests. Most of these applications involve laser induced fluorescence (LIF) of dye tagged structures from whole cells down to single DNA bases. In these techniques, the biological sample is treated with a strong fluorescent dye or chemical that has an affinity for a specific chemical structure, environment or site within a cell or ensemble of cells. The sample is then irradiated with the laser light (Fig. 2), and the dye tag fluoresces, thus indicating its physical location.

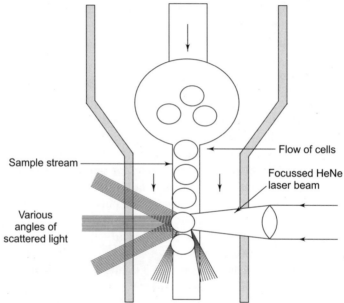

Fig. 2   Cell sizing and sorting with the help of HeNe laser.

Medical/Surgical Benefits

A growing application for helium neon lasers, particularly in the green and yellow, is an alignment tool for other lasers, such as CO_2 and Nd:YAG infrared lasers used in surgical procedures. The visible beam of the HeNe is coaligned (Fig. 3), through the beam delivery optics so that the surgeon can accurately direct the high power infrared beam, before it is switched on. In addition to pointing stability, this application relies on the visibility/contrast of the HeNe beam on human tissue. Early systems used the red HeNe, but a red spot is often difficult to see on tissue, particulary when the tissue is highly vascularized or when free blood is present. Now this application has switched virtually exclusively to green and yellow laser which offer much better eye contrast.

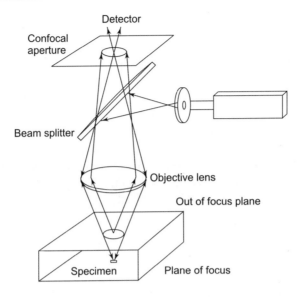

Fig. 3   Use of HeNe Laser for alignment in medical/surgical operations.

High Speed Printing

Helium neon lasers have long been used in the graphic arts for both printing and scanning. As with any laser scanner, the principle is simple; the focused laser spot is swept in one axis while the image, film or printer drum moves in the other axis to deliver a xy raster scan. Use of a small laser spot enables high spatial resolution.

Now higher power (10 to 40 mW) red lasers are finding use in a new medical application of this type, namely to enable digitization, storage and reprinting of x-ray images. To avoid the need for storing large volumes of developed x-ray films, medical facilities are using HeNe based scanners to digitize the information.

Interferometric Metrology and Velocimetry

In the area of industrial alignment, another important group of applications utilizes the coherence of the helium neon laser beam. These techniques are based on interference and use the wave periodicity of the light beam as a very fine ruler. The position of an object in the path of the beam is computed from the phase of the light reflected from it. Interference between the object beam and a reference beam (split-off from the same laser) provides measurable intensity variations which yield this phase information. This technique allows both the position and velocity of an object or surface to be determined with extraordinary accuracy.

Barcode Reading

Automated reading of labels using bar codes has revolutionized the way industrial products are routed and streamlined the job of the supermarket checker. Helium neon remains the dominant laser type in both these application areas. In industrial systems, the laser barcode reader provides a low cost method of positively identifying, sorting, and controlling inventories of all types of products. Food, commercial, and industrial warehouses and distribution centers depend on accurate unit identification and control for efficient and cost-effective operation. In these applications, the reader is positioned above or beside a conveyor system running at speeds of about 400 to 600 ft. per min. The articles are passed by the reader, the barcodes are scanned, and the identified items are then routed to their proper destinations.

Entertainment

Red helium neon lasers have been used for entertainment purposes for many years because they offer bright visible laser light at a very affordable cost. The development of reliable commercial HeNe lasers at other wavelengths has naturally led to an increased usage of HeNe lasers in this field. Examples include indoor laser shows, war games (laser tagging) and of course holographic displays.

Acknowledgements

The author is thankful to Prof. J. Behari, Head, Department of Environment Studies, JNU, Delhi, for presenting this article. Thanks are also due to Mr. K.S. Jindal, Director, LASTEC for his guidance and encouragement.

Topics in Electromagnetic Waves: Devices, Effects and Applications
Edited by J. Behari
Copyright © 2005, Anamaya Publishers, New Delhi, India

19. Millimeter-Waves

G.P. Srivastava

Department of Electronic Science, University of Delhi, South Campus,
New Delhi-110 021, India

Abstract: With excessive use of microwaves for various types of communication, no frequency is available for further growth. The present article describes the use of millimeter-waves for communication as well as many other uses. A brief description of activities in various countries is also discussed.

Introduction

What are millimeter-waves? It is difficult to define boundaries. Starting from 30 GHz to 300 GHz (1 mm wavelength to 0.1 mm wavelength) is generally referred to as millimeter-waves. At present many applications in this region are found.

Jagdish Chandra Bose succeeded in generating electromagnetic waves (EW) with wavelength of 5 millimeters and coined the word *millimeter-waves*. His radiator was composed of two hollow hemisphere interposed with a small sphere and generated Radio Waves through electric discharge. These electromagnetic waves were radiated out through a device now called *horn antenna*. He used a galena crystal as receiver, which was later used as detector crystal radio sets and succeeded in demonstrating reflection, refraction and total internal reflection; double refraction and polarization by crystal of tourmaline and nemalite of these electromagnetic waves and this proved their identity with light waves. As early as 1895, he gave a public demonstration in Kolkata of remote control devices operated by millimeter-waves. He succeeded in operating remote relay and igniting gun-powder kept three rooms away.

Advantages of Using Millimeter-Waves for Communication

Following are the advantages of using millimeter-waves for communications:

1. *Small Size:* Because of short wavelength compared to Microwaves, millimeter-wave components are smaller in size. It helps in building small-size system.
2. *Wide Bandwidth:* There are transmission windows at 35, 94, 140 and 220 GHz where large bandwidths of the order of 16, 23, 26 and 70 GHz are available for communication. Thus, many individual frequencies can be used to avoid interference and achieve a high level of electromagnetic compatibility.

3. *Narrow Beam Width:* At millimeter wavelengths, smaller radiated beams can be obtained for a given antenna size. For example, a 12 cm diameter antenna provides 1-8° beam width at 94 GHz compared to 18 at 9.4 GHz. Narrow Beam Width provides higher resolution and hence improved accuracy. It also minimizes losses and noise due to side lobe returns Narrow Beam Width makes jamming difficult.
4. Millimeter-Waves Systems can function in smoke, dust, fog and haze as well as in the night time.

Limitations

1. Component costs are relatively high.
2. Reliability and availability of components.
3. Operation range is relatively short even under clear atmospheric conditions. These may be due to equipment limitations, but primarily due to absorption in water vapour and oxygen molecules.
4. Range is reduced under certain absorption conditions due to scattering and absorption in rains. First two of these have been considerably improved.

Apart from the unique propagation behaviour millimeter-wave applications take advantage of highly directive nature of propagation beam. Thus, millimeter-wave systems are small in size and light in weight compared to microwave counterparts. The lack of availability of frequencies in microwave bands commonly used is an important reason to move into high frequency region.

It is only today that the *demand pull* has taken over from usual technological push. The spreading applications of millimeter-wave today have become possible due to rapidly advancing component technology of millimeter-waves.

Millimeter-Wave Circuit Elements

There has recently been rapid progress in manufacturing technology of Monolithic Microwave/Millimeter Wave Monolithic Integrated Circuit (MMIC). MMICs have intrinsically many practical advantages in terms of cost, performance and reliability.

If dielectric waveguides such as image lines are used, the loss can be substantially reduced, but radiation at the bends and discontinuities in dielectric waveguides may cause some trouble. Yoneyama proposed NRD-waveguide (Non-radiative dielectric waveguide). It suppresses radiation completely. Passive circuit components such as bends, directional couplers, filters and resonators have been fabricated.

Hybrid integrated circuits employing beam led Schottky barrier GaAs diode on fin lines or micro-strip circuits in production for frequencies up to 150 GHz. Noise figures of 7 db are typical at 94 GHz.

Millimeter-Wave Sources

Many millimeter-wave sources are being developed. Three groups are as follows:
1. Gyratron
  (i) Gyro-Monotron Oscillator
  (ii) Gyro-TWT
  (iii) Gyro-Klystron
2. Ubitron (An Undulated Beam Interaction Electron Device)
    Like conventional travelling wave tube, it is an O-type travelling-wave amplifier and can produce peak power of 1-60 mW.
3. Peniotron
    The interaction between EM wave propagating in a double pair ridged waveguide and a thin hollow electron beam around flux lines of an applied dc magnetic field at cyclotron angular frequency ω_c and drifting parallel to the field at a constant velocity v_{11}. By considering the Doppler Shift and optimum interaction conditions we get

$$\omega - \beta \, v_{11} = 2p \, \omega_c$$

where ω and β are angular frequency and wave number, respectively, and p a positive integer.

Millimeter Semiconductor Sources

GaAs FETs can be used at pretty high frequency but beyond 60 or 70 GHz special devices are used. One of them is HEMT (High Electron Mobility Transistor). It is a low noise, high-speed transistor, utilizes two dimensional electron gas which is a layer of electron cloud formed at hetero junction interface. Dr. R. Esaki (1970) proposed a multilayer super-lattice consisting of multilayer of two semiconductors with different band gaps such as GaAs and AlGaAs.

    The Resonant Tunellin of Diode (RTD) can be used for generation of highest frequency of 700 GHz. The resonant tunnelling process and other electron wave interference can be controlled by electrical signals to create novel quantum interference device. BRINT (Bragg Reflection and Interference Transistor) is an example of the same. In it a periodic potential is introduced in FET channels so that electron wave of specific wavelengths undergo the Bragg reflection, resonant tunnelling of diodes provide excellent negative resistance characteristics. Since RTD is determined by the energy ΔE (or the bandwidth) of transmission filter characteristics through the uncertainty principle $h/\Delta E$.

    Remarkable developments in semiconductor technology have allowed preparation of variety of Layered Micro Structure (LMS) with thickness controllable to nearly one atomic layer. Electron traversing through or confined in such structure exhibit unique and unpredicted properties since quantum wave nature of electron manifest themselves particularly in their motion along the staking direction.

Nishizawe (1958) proposed TUNNETT in a two terminal negative resistance diode using Tunnel effect inverse biased p^+-n^+-n-n^+ diode operates at high frequency (THz), has low noise and operates at low biased voltage. The oscillator frequency is given by

$$f_{osc} = 3\ V_s/4\ w$$

where V_s is the saturation velocity of an electron and w is the thickness of the drift region.

Its operation in T-band (110-170 GHz), Y-band (170-210 GHz) and D-band (220-325 GHz) has already been demonstrated.

Some Applications of Millimeter-Waves

Some of the numerous applications are discussed as follows:

1. Millimeter-Wave Radar

Millimeter-wave radar has
  (i) High antenna gain with small aperture
 (ii) High track and guidance accuracy
(iii) Reduced ECM vulnerability
 (iv) Operational low elevation angles capability without significant multi path and ground clutter interference
  (v) Multi path target discrimination capability
 (vi) Mapping quality resolution
Millimeter-wave radars operating at short-to-moderate ranges are quite attractive for surveillance and target acquisition. These are especially effective against two types of targets: (i) high altitude borne targets and (ii) ground targets, where it is necessary to reduce clutter return.

2. Millimeter-Wave Radiometers

  (i) Remote sensing applications
 (ii) Detecting capability of noise temperature as seen by antenna
(iii) Capability of guiding missiles
 (iv) Capability of applying it in Radio Astronomy
Other radiometric applications include thermography or trilogical imaging which has shown the possibilities of detection of cancerous tumor in human bodies. It can also be used for therapy of cancerous growth by focusing the millimeter-waves on the target.

3. Millimeter-Waves Communication System

Perhaps millimeter-waves are most widely used for communication purpose. For covering route lengths over 2000 km, digital radio and satellite networks are expected to dominate. This is in contrast to situation earlier when analog transmission system utilizes *rf* transmission bandwidth. What makes digital system competitive was development of low-cost integrated digital circuitry. This makes it possible to have regenerative repeater interfacing and multiplexing hardware at reduced cost.

Commercial applications of millimeter-waves are being more and more exploited. This has been helped by advent of low-cost integration procedure, i.e. hybrid and monolithic integration techniques. Having the necessary technology at hand, mature and commercially exploitable techniques are available. From indoor high-speed radio communication systems at 10.6 GHz as proposed by NTT (Japan), our "gateways" for mobile or new deregulated stationary systems at 38 GHz and 55 GHz, being available from Northern Telecom (UK) and WLAN's at 60 GHz like MBS project being under research within European RACE-programme, to optic microwave hybrid approach for Pico Cellular PCN's as proposed by Alcatel-Sel of Germany and CNET of France.

Digital systems are high noise and distortion resistive. Error detection and correction are possible. In USA the use of Time Division Multiplexed (TDM) short-haul cable system initiated the new development of digital systems. These use a basic group of 24 digitized multiplexed voice channels having a data rate of 1.544 Mb/s. The availability of cheap hardware interfacing A/D converter each voice channel digitally coded and transformed to the next level of hierarchy. The European system based on hierarchy known as CCIT gives small SNR. A heterodyne transreceiver is shown in Fig. 1.

QPSK modulation is generally used. It can be MPSK. In QPSK modulation, two carriers phase shifted by 90°, are modulated by separate data streams. With ideal components the two modulated signals are fully independent, and therefore, double the capacity without increasing the bandwidth. If arbitrary length of phase vector of modulated signal is allowed an even greater degree of freedom is achieved than constant amplitude. The resulting modulation scheme is then defined as an M-QAM (Quadraline Amplitude Modulation with M states of Phase Vector).

Cellular Mobile Communication

Of all tremendous advances in data communications and telecommunications perhaps the most revolutionary is development of cellular networks. The essence of using cellular network is to use multiple low-power transmitters of the order of 100 W or less. As the range of such transmitters is small, the area is divided into cells served by its own antenna. Each cell is allotted band of frequencies as is served by a base station consisting of transmitter, receiver and central unit. Each cell adjacent are assigned different frequencies to avoid interference. A hexagon is the shape of each cell so that no space is left uncovered.

Since 1990 commercial Mobile Satellite Services (MSS) have been used in several countries. Currently MSS systems use L- and S-band frequencies. For future SS expansion millimeter-wave region is being investigated.

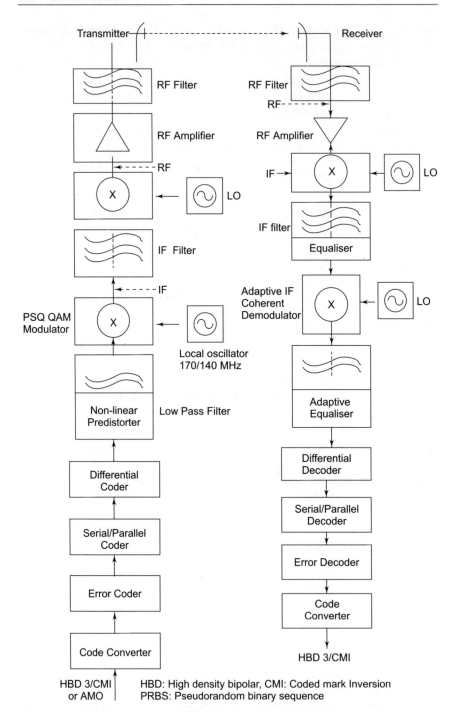

Fig. 1   A heterodyne transreceiver.

NASA is already making use of Advanced Communication Technology (ACTs) satellite. Japan has launched Communications and Broadcasting Test Satellite (COMSETS), which provides a test bed for evaluating future, advanced mobile and personal satellites using millimeter-waves.

Wireless Local Loop (WLL)

Wireless local loop (WLL) is throwing a challenge to cellular communications. WLL alternatives are narrow band, which offer a replacement for the existing telephone exchange and are broadband which provide high-speed two-way voice and data service.

For most high-speed WLL schemes frequencies in millimeter region are used. The reasons for using frequencies in this region are:
1. Wide unused frequency bands (~ 25 GHz) are available.
2. At these high frequencies wide channel bandwidths can be used, providing high data rates.
3. Small size trans-receivers and antenna array can be used.

Automotive Collosion Avoidance Radar

Small size antenna can fit into the front of a car. Such radars are built for short-range use only, up to 200 m maximum, under adverse weather conditions. The Autonomous Intelligent Cruise Control (AICC) radar checks approach. The technical realization of AICC needed is antenna scanning. This should be done electronically by sequential switching employing PIN diodes. The FM-CW censor has output power of 15 W. A 94 GHz radar was developed by Philips Microwave of UK for this purpose. If these ideas are extended to railway system, collision in bad weather can be avoided.

Road Transport Informatics

In 1980 vehicle-to-roadside communication was tested. Fujitsu is developing fully monolithic 60 GHz RF chip for this purpose. The 61.5 GHz vehicle sensors based on Doppler Principle have been successfully applied for traffic jam avoidance or warning. Due to high frequency small antenna dimensions allow good focusing so that only desired lane is illuminated. Short-range applications are not hampered by high atmospheric absorption. Economic reuse of frequencies due to high atmospheric absorption is feasible.

Skynetwork (Access Network Using Stratospheric Platform)

This has a potential to become the third communication infrastructure after terrestrial and satellite communications. The platforms keep their positions at about 20 km high in stratosphere. Each one covers an area of 40 km × 40 km. An access link is the link between platform station and user station. The frequency band of access link is expected to use a mm wave band.

A 600 MHz bandwidth in a 48/47 Hz ban has been allocated for fixed services of high altitude station.

Conclusion

Millimeter-wave technology can meet outstanding system requirements. Short-haul communications is a reality now. Vehicle-to-roadside, vehicle-to-vehicle communication is possible today at 60 GHz. The 77 GHz AICC radar sensors will make driving more comfortable and safer. A successful innovation process in millimeter-wave has been established. It is the market drive which has become most important factor for making new applications. The emphasis of millimeter-wave application now has shifted from military to commercial. It is now the consumer who will decide newer applications of millimeter-waves.

Index